Physics in a New Era

....

An Overview

Physics Survey Overview Committee
Board on Physics and Astronomy
Division on Engineering and Physical Sciences
National Research Council

NATIONAL ACADEMY PRESS
Washington, D.C.

This project has been supported by the National Aeronautics and Space Administration under Grant No. NAG 5-6839, the Department of Energy under Contract No. DE-FG02-98ER-41069, and the National Science Foundation under Grant No. PHY-972-4780. Any opinions, findings, and conclusions or recommendations expressed in this material are those of the author(s) and do not necessarily reflect the views of the sponsors.

Front cover: An example of a form of abstract composition known as Marian vectors, based on mathematical processes. Courtesy of the National Center for Supercomputing Applications at the University of Illinois at Urbana-Champaign.

International Standard Book Number 0-309-07342-1
Library of Congress Catalog Card Number 2001-089190

Copies of this report are available from:
    National Academy Press
    2101 Constitution Avenue, N.W.
    Lockbox 285
    Washington, DC 20055
    (800) 624-6242 or (202) 334-3313 (in the Washington metropolitan area)
    Internet <http://www.nap.edu>
and
    Board on Physics and Astronomy
    National Research Council, HA 562
    2101 Constitution Avenue, N.W.
    Washington, DC 20418

Printed in the United States of America

THE NATIONAL ACADEMIES

National Academy of Sciences
National Academy of Engineering
Institute of Medicine
National Research Council

The National Academy of Sciences is a private, nonprofit, self-perpetuating society of distinguished scholars engaged in scientific and engineering research, dedicated to the furtherance of science and technology and to their use for the general welfare. Upon the authority of the charter granted to it by Congress in 1863, the Academy has a mandate that requires it to advise the federal government on scientific and technical matters. Dr. Bruce Alberts is president of the National Academy of Sciences.

The National Academy of Engineering was established in 1964, under the charter of the National Academy of Sciences, as a parallel organization of outstanding engineers. It is autonomous in its administration and in the selection of its members, sharing with the National Academy of Sciences the responsibility for advising the federal government. The National Academy of Engineering also sponsors engineering programs aimed at meeting national needs, encourages education and research, and recognizes the superior achievements of engineers. Dr. William A. Wulf is president of the National Academy of Engineering.

The Institute of Medicine was established in 1970 by the National Academy of Sciences to secure the services of eminent members of appropriate professions in the examination of policy matters pertaining to the health of the public. The Institute acts under the responsibility given to the National Academy of Sciences by its congressional charter to be an advisor to the federal government and, upon its own initiative, to identify issues of medical care, research, and education. Dr. Kenneth I. Shine is president of the Institute of Medicine.

The National Research Council was established by the National Academy of Sciences in 1916 to associate the broad community of science and technology with the Academy's purposes of furthering knowledge and advising the federal government. Functioning in accordance with general policies determined by the Academy, the Council has become the principal operating agency of both the National Academy of Sciences and the National Academy of Engineering in providing services to the government, the public, and the scientific and engineering communities. The Council is administered jointly by both Academies and the Institute of Medicine. Dr. Bruce Alberts and Dr. William A. Wulf are chairman and vice chairman, respectively, of the National Research Council.

PHYSICS SURVEY OVERVIEW COMMITTEE

Preface

Physics in a New Era: An Overview is the culmination of the National Research Council survey series *Physics in a New Era*. The survey was proposed by the Board on Physics and Astronomy, continuing the tradition of periodic reviews of physics by the National Research Council. The overview is the final volume of the survey and was welcomed and supported by the Department of Energy, the National Science Foundation, and the National Aeronautics and Space Administration. Volumes published previously in the series are *Atomic, Molecular, and Optical Science: An Investment in the Future* (1994) (the AMO science survey), *Plasma Science: From Fundamental Research to Technological Applications* (1995), *Elementary-Particle Physics: Revealing the Secrets of Energy and Matter* (1998), *Nuclear Physics: The Core of Matter, The Fuel of Stars* (1999), *Condensed-Matter and Materials Physics: Basic Research for Tomorrow's Technology* (1999), and *Gravitational Physics: Exploring the Structure of Space and Time* (1999). In addition to these six volumes, which are known as the area volumes, the survey includes four more: *Cosmology: A Research Briefing* (1995), *Cosmic Rays: Physics and Astrophysics* (1995), *Neutrino Astrophysics: A Research Briefing* (1995), and *The Physics of Materials: How Science Improves Our Lives* (1997). A related study that was recommended by the AMO science study is entitled *Harnessing Light: Optical Science and Engineering for the 21st Century* (1998).

The area volumes review recent achievements, describe goals of the subdisciplines for the new decade, and identify the research programs with the highest priority for advancing those goals. The six area volumes are available online through the Board on Physics and Astronomy's Web site, <http://www.national-academies.org/bpa/reports>. Since each volume surveys a rapidly developing area, the later volumes are naturally more up to date than those completed several years ago. The AMO science study is already being updated. The recommendations, nevertheless, remain pertinent and have served as a foundation for the present volume, which addresses physics as a whole.

The Physics Survey Overview Committee was asked to survey the field of physics broadly, identify priorities, and formulate recommendations, complementing the field-specific discussions in the area volumes. The overview assesses the state of physics in four broad categories—quantum manipulation and new materials, complex systems, structure and evolution of the universe, and fundamental laws and symmetries—emphasizing the unity of the field and the strong commonality that links the different areas, while highlighting new and emerging ones. The importance of international cooperation in many areas of physics is emphasized. The overview goes on to discuss the challenges facing physics education, from K-12 through graduate school, and the expanding connections of physics with other fields of engineering and science, including the biological sciences. It also describes the impact of physics on the economy, in particular on the development of information technology; the role of physics in national security; and the many contributions of physics to health care.

The breadth of the overview is reflected in its priorities and recommendations. They are meant to sustain and strengthen all of physics in the United States and enable the field to serve important national needs. They are not subfield-specific, but the committee believes that they are compatible with and complementary to the priorities and recommendations of the area volumes. The report identifies six high-priority arenas of research, cutting across the traditional subfields. It concludes with nine recommendations touching on levels of support, education, national security, planning and organization, and the role of information technology in physics.

Acknowledgments

The committee was helped in its work by a great many people. It is especially grateful to Bertram Batlogg, Mark Brandon, D. Allan Bromley, Rad Byerly, Sidney Drell, Murray Gibson, Steven Girvin, Will Happer, Mark Ketchen, Steven Koonin, James Langer, Thomas Mason, Jeffrey Park, Nicholas Samios, F.M. Scherer, Robert Socolow, and Peter Webster. It also expresses its gratitude to the American Physical Society, to the Society's executive officer, Judy Franz, and to the many members of the APS who responded so thoughtfully to its request for advice.

The committee would like to thank Donald C. Shapero, Robert L. Riemer, Achilles Speliotopoulos, and the entire staff of the Board on Physics and Astronomy for their many valuable contributions throughout the preparation of the overview.

Grant support for the work of the committee has come from the National Science Foundation, the Department of Energy, and the National Aeronautics and Space Administration. The committee thanks them for this support. Finally, it acknowledges its great debt to David N. Schramm, under whose chairmanship of the Board on Physics and Astronomy the decadal survey *Physics in a New Era* began. The committee dedicates this overview to his memory.

Thomas Appelquist, *Chair*
Physics Survey Overview Committee

Acknowledgment of Reviewers

This report has been reviewed in draft form by individuals chosen for their diverse perspectives and technical expertise, in accordance with procedures approved by the National Research Council's Report Review Committee. The purpose of this independent review is to provide candid and critical comments that will assist the institution in making its published report as sound as possible and to ensure that the report meets institutional standards for objectivity, evidence, and responsiveness to the study charge. The review comments and draft manuscript remain confidential to protect the integrity of the deliberative process. We wish to thank the following individuals for their review of this report:

John A. Armstrong, IBM Corporation (retired),
Gordon Baym, University of Illinois at Urbana-Champaign,
Radford Byerly, Independent Consultant,
Persis Drell, Cornell University,
David Gross, University of California at Santa Barbara,
Sol Gruner, Cornell University,
William Happer, Princeton University,
Daniel Kleppner, Massachusetts Institute of Technology,
Carl Lineberger, JILA/University of Colorado,
John C. Mather, NASA Goddard Space Flight Center,
Albert Narath, Lockheed Martin Corporation (retired),
Venkatesh Narayanamurti, Harvard University,
V. Adrian Parsegian, National Institutes of Health,
Julia Phillips, Sandia National Laboratories,
Judith Pipher, University of Rochester, and
Paul Steinhardt, Princeton University.

Although the reviewers listed above have provided many constructive comments and suggestions, they were not asked to endorse the conclusions or recommendations, nor did they see the final draft of the report before its release. The review of this report was overseen by Pierre Hohenberg, Yale University, appointed by the Report Review Committee, who was responsible for making certain that an independent examination of the report was carried out in accordance with institutional procedures and that all review comments were carefully considered. Responsibility for the final content of this report rests entirely with the authoring committee and the institution.

Contents

List of Sidebars

Physics in a New Era

An Overview

Executive Summary

The advances and breakthroughs of 20th-century physics have enriched all the sciences and opened a new era of discovery. They have touched nearly every part of our society, from health care to national security to our understanding of Earth's environment. They have led us into the information age and fueled broad technological and economic development. The pace of discovery in physics has quickened over the past two decades. New microscopic devices are being developed with a host of potential applications, and instruments of unprecedented sensitivity and reach are being created and employed. Physics at the tiniest distances is being linked to the origin and fate of the universe itself.

PHYSICS FRONTIERS

A second quantum revolution is under way. Physicists exploring and controlling the properties of collections of atoms are shrinking the materials they study to sizes at which quantum properties play a key role. The study of this nanoscale regime, smaller than the wavelength of visible light, is ushering in an era of powerful electronic devices.

At the same time, as they study ever more complex systems, physicists are joining forces with biologists to understand life and with geologists to explore Earth and the planets. Dramatic advances in computing are responsible for much of this progress, allowing vast amounts of data to be collected and understood and enabling many of the most complex phenomena encountered in nature to be analyzed numerically.

In astrophysics and cosmology, a new generation of space-based and Earth-based instruments has brought about a golden age. Exciting questions are being addressed: Is the expansion of the universe today accelerating as a result of some mysterious form of energy? Did the universe undergo a period of very rapid expansion (inflation) at its earliest moments? How do black holes form?

Amazingly, these questions about the cosmos are being linked directly to physics at the tiniest distances. The exploration of the next high-energy frontier at a new generation of particle colliders will illuminate the origins of elementary particle masses and may reveal a profound unification of all the forces of nature.

PHYSICS AND SOCIETY

With physics now connected strongly to the other sciences and contributing to many national needs, education in physics is of vital importance. Physics is at the heart of the technology driving our economy, and broad scientific literacy must be a primary goal of physics education at all levels. To achieve this goal, to provide an education linked to the wider world that is so important for members of the high-tech work force, and to draw more students into careers in science will require the best efforts of university physics departments and national laboratories.

It is an international society that physics and physics education must reach in this new era. The problems that physics can address are global problems, and physics itself is becoming a more international enterprise. New modes of international cooperation must be created to plan and operate the large facilities that are increasingly important for frontier research.

SCIENTIFIC PRIORITIES AND OPPORTUNITIES

The accomplishments of physics, the growing power of its instruments, and its expanding reach into the other sciences have generated an unprecedented set of scientific opportunities. The committee has identified six such "grand challenges," listed below in no particular order. They range across all of physics, extending from purely theoretical work and numerical simulation to research requiring large experimental facilities. They are selective: Some coincide with the priorities set forth in the area volumes,[1] while others cut more broadly across the whole of physics, overlap other areas of science, or are of growing importance for technology. The committee chose them based on their intrinsic scientific importance, their potential for broad impact and application, and their promise for major progress

[1]See the preface for a list of the area volumes and the Web site address through which they can be accessed online.

during the next decade. It urges that these high-priority areas be supported strongly by universities, industry, the federal government, and others in the years ahead.

Developing Quantum Technologies

The ability to manipulate individual atoms and molecules will lead to new quantum technologies with applications ranging from the development of new materials to the analysis of the human genome. This ability allows the direct engineering of quantum probabilities, producing novel phenomena such as the presence of many atoms in the same quantum mechanical state with a high probability of spatial overlap and entanglement. Quantum overlap can sometimes extend over distances very large compared to a single atom, as in gaseous Bose-Einstein condensates. A new generation of technology will be developed with construction and operation entirely at the quantum level. Measurement instruments of extraordinary sensitivity, quantum computation, quantum cryptography, and quantum-controlled chemistry are likely possibilities.

Understanding Complex Systems

Theoretical advances and large-scale computer modeling will enable phenomena as complicated as the explosive death of stars and the properties of complex materials to be understood at a depth unavailable only a few years ago. The rapid advances of massively parallel computing, coupled with equally impressive developments in theoretical analysis, have generated an extraordinary growth in our ability to model and predict complex and nonlinear phenomena and to visualize the results. Problems that may soon be rendered tractable include the strong nuclear force, turbulence and other nonlinear phenomena in fluids and plasmas, the origin of large-scale structure in the universe, and a variety of quantum many-body challenges in condensed-matter, nuclear, atomic, and biological systems. The study of complex systems is inherently of great breadth: Improvements in the understanding of radiation transport, for example, will advance both astrophysics and cancer therapy.

Applying Physics to Biology

Because all essential biological mechanisms ultimately depend on physical interactions between molecules, physics lies at the heart of the most profound insights into biology. Problems central to biology such as the way molecular chains fold to yield the specific biological properties of proteins will become accessible to analysis through basic physical laws. Current challenges include the biophysics of cellular electrical activity underlying the functioning of the nervous system, the circulatory system, and the respiratory system; the biomechanics of the motors responsible for all biological movement; and the mechanical and electrical properties of DNA and the enzymes essential for cell division and all cellular processes. Tools developed in physics, particularly for the understanding of highly complex systems, are vital for progress in all these areas. Theoretical approaches developed in physics are being used to understand bioinformatics, biochemical and genetic networks, and computation by the brain.

Creating New Materials

Novel materials will be discovered, understood, and employed widely in science and technology. The discovery of materials such as high-temperature superconductors and new crystalline structures has stimulated new theoretical understanding and led to applications in technology. Several themes and challenges are apparent—the synthesis, processing, and understanding of complex materials composed of more and more elements; the role of molecular geometry and motion in only one or two dimensions; the incorporation of new materials and structures in existing technologies; the development of new techniques for materials synthesis, in which biological processes such as self-assembly can be mimicked; and the control of a variety of poorly understood, nonequilibrium processes (e.g. turbulence, cracks, and adhesion) that affect material properties on scales ranging from the atomic to the macroscopic.

Exploring the Universe

New instruments through which stars, galaxies, dark matter, and the Big Bang can be studied in unprecedented detail will revolutionize our understanding of the universe, its origin, and its destiny. The universe itself is now a laboratory for the exploration of fundamental physics: Recent discoveries have strengthened the connections between the basic forces of nature and the structure and evolution of the universe. New measurements will test the foundations of cosmology and help determine the nature of dark matter and dark energy, which make up 95 percent of the mass-energy of the universe. Gravitational waves may be directly detected, and the predictions of

Einstein's theory for the structure of black holes may be checked against data for the first time. Questions such as the origin of the chemical elements and the nature of extremely energetic cosmic accelerators will be understood more deeply. All of this has given birth to a rich new interplay of physics and astronomy.

Unifying the Forces of Nature

Experiment and theory together will provide a new understanding of the basic constituents of matter. The mystery of the nature of elementary particles deepened in the 1990s with the discovery of the extraordinarily heavy top quark and the observation of oscillations in neutrinos from the Sun and the upper atmosphere, suggesting that neutrinos have extremely tiny masses. During the next decade the unknown physics responsible for elementary particle masses and other properties will begin to reveal itself in experiments at a new generation of high-energy colliders. Possibilities range from the discovery of new and unique elementary particles to more exotic scenarios involving fundamental changes in our description of space and time.

Determining this new physics is an important step toward an historic goal: the discovery of a unified theoretical description of all the fundamental forces of nature—the strong nuclear force, the electroweak forces, and gravity. The most promising and exciting framework for unifying gravity with the other forces is string theory, which proposes that all elementary particles behave like strings at very tiny distances. String theory has also given birth to new and vibrant intersections between physics and pure mathematics. The next decade will see much progress toward the goal of discovering a unified theory of the forces of nature.

RECOMMENDATIONS

From this survey of physics and its broad impact and the identification of six high-priority scientific opportunities, the committee has developed a set of nine recommendations. They are designed to strengthen all of physics and to ensure the continued international leadership of the United States. They address the support of physics by the federal government and the scientific community; physics education; the role of basic physics research in national security; the increasingly important role of partnerships among universities, industry, and national laboratories; the stewardship of federal science agencies; and the rapidly changing role of information technology in physics research and education.

Recommendation 1: Investing in Physics. To allow physics to contribute strongly to areas of national need, the federal government and the physics community should develop and implement a strategy for long-term investment in basic physics research. Key considerations in this process should include the overall level of this investment necessary to maintain strong economic growth driven by new physics-based technologies, the needs of other sciences that draw heavily on advances in physics, the expanding scientific opportunities in physics itself, the cost-effectiveness of stable funding for research projects, the characteristic time interval between the investment in basic research and its beneficial impact, and the advantages of diverse funding sources. The Physics Survey Overview Committee believes that to support strong economic growth and provide essential tools and methods for the biomedical sciences in the decade ahead, the federal investment in basic physics research relative to GDP should be restored to the levels of the early 1980s.

Recommendation 2: Physics Education. Physics departments should review and revise their curricula to ensure that they are engaging and effective for a wide range of students and that they make connections to other important areas of science and technology. The principal goals of this revision should be (1) to make physics education do a better job of contributing to the scientific literacy of the general public and the training of the technical workforce and (2) to reverse, through a better-conceived, more outward-looking curriculum, the long-term decline in the numbers of U.S. undergraduate and graduate students studying physics. Greater emphasis should also be placed on improving the preparation of K-12 science teachers.

Recommendation 3: Small Groups and Single Investigators. Federal science agencies should assign a high priority to providing adequate and stable support for small groups and single investigators working at the cutting edge of physics and related disciplines.

Recommendation 4: Large Facilities and International Collaboration. While planning and priority setting are important for all of physics, they are especially critical when large facilities and collaborations are necessary. To plan successfully, the community of physicists in the United States and abroad must develop a broadly shared vision and communicate this vision clearly and persuasively. Planning and implementation for the very largest facilities should be international. The federal government should develop effective mechanisms for U.S. participation and leadership in international scientific projects, including clear criteria for entrance and exit.

Recommendation 5: National Security. Congress and the Department of Energy should ensure the continued scientific excellence of the Department of Energy's Office of Defense Programs' national laboratories by reestablishing the high priority of long-term basic research in physics and other core competencies important to laboratory missions.

Recommendation 6: Partnerships. The federal government, universities and their physics departments, and industry should encourage mutual interactions and partnerships, including industrial liaison programs with universities and national laboratories; visitor programs and adjunct faculty appointments in universities; and university and national laboratory internships and sabbaticals in industry. The federal government should support these programs by helping to develop protocols for intellectual property issues in cooperative research.

Recommendation 7: Federal Science Agencies. The federal government should assign a high priority to the broad support of core physics research, providing a healthy balance with special initiatives in focused research directions. Federal science agencies should continue to ensure a foundation that is diverse, evolving, and supportive of promising and creative research.

Recommendation 8: Peer Review. The peer review advisory process for the allocation of federal government support for scientific research has served our nation well over many decades and is a model worldwide for government investment in research. The peer review process should be maintained as the principal factor in determining how federal research funds are awarded.

Recommendation 9: Physics Information. The federal government, together with the physics community, should develop a coordinated approach for the support of bibliographic and experimental databases and data-mining tools. The use of open standards to foster mutual compatibility of all databases should be stressed. Physicists should be encouraged to make use of these information technology tools for education as well as research. The bibliographic archive based at Los Alamos National Laboratory has played an important role and it should continue to be supported.

Introduction

Physics matters because it stands where wonder about the workings of the world meets our most practical demands. Like quicksilver, physics darts this way and that through the tangle of disciplines, making connections, building instruments, explaining why things work. Who would have expected that Maxwell's theory of electricity and magnetism, born in the midst of Victorian Cambridge, would a few decades later link continents by radio? That Einstein's relativity, launched on a wooden podium in a Swiss patent office, would astonish the world with radical new ideas of time and space and then be key to the understanding of fission, nuclear power stations, and weapons? Or that the sharing of data necessitated by the inquiry into quark dynamics in particle colliders would lead to the establishment of the World Wide Web?

Physics matters because issues of understanding and practicality rarely stay apart for long. In 1947, Bell Laboratories physicists working on a new amplifier built the first prototype transistor; subsequent exploration of semiconductors led to a dramatically new atomic understanding of condensed matter physics. That new atomic understanding, in turn, fed back into the massive industry that has placed microelectronics at the center of today's economy. Techniques that physicists developed as they grappled with the theory of magnetism aided in the understanding of superconductivity and then migrated to elementary particle physics, where they led to a new grasp of the unity of the basic forces that hold matter together.

Where is physics? It is not now and never was isolated in university departments. A generation ago, physicists made common cause with electrical engineers to build the radar facilities that played such a key role in World War II; they partnered with industrial engineers, chemical engineers, and metallurgists to construct the vast nuclear plants that have produced both electricity and weapons. More recently, physicists have been entering into new partnerships with industry as they construct the infinitesimal circuits and machines of nanotechnology. They have joined with biologists: Perhaps in the not-too-distant future they will use optical tweezers to rearrange the genetic code. They have begun to explore radical new ideas of quantum computing. Physicists are to be found in NASA, in industries, and, increasingly, on Wall Street as techniques born in the study of physical systems begin to play an important role in examining the dynamics of stocks, bonds, options, and hedge funds. Physics is large in the vast national laboratories of Los Alamos, Oak Ridge, Fermilab, the Stanford Linear Accelerator Center (SLAC), and Brookhaven. Physics is small in the start-up optics company or in the bench-top experiments that are reconfiguring the study of new materials, optical phenomena, biophysics, and magnetic media.

Precisely because physics is everywhere, from computer printers, copying machines, and laser-driven checkout counters to precision weapons, surgical instruments, airplane surfaces, and medical diagnostics, it is a tall order to survey the whole of it. One way to cast a first glance over the landscape is to think of the different distance scales of nature.

A few years ago, Sebastian Junger wrote *A Perfect Storm*, a moving account of a few small boats caught in the harsh fall Atlantic during the extraordinarily violent storm of October 1991. We learn of the killer waves, some almost a hundred feet high and roughly that long, that threatened the lives of anyone so unfortunate as to be at sea. Any small craft would eventually lose the power to face into the waves and once caught sideways would almost certainly be thrown keel up and sunk. Even on a calm day, the complexities of the sea seem impossible to understand, especially on discovering that there are waves of every conceivable size, from a fraction of an inch to giant ocean surges that stretch for a thousand miles. But there is a simplification available, one that in some ways undergirds the entirety of physics. It is this: Not all the ups and downs of a boat at sea are equally threatening. Waves much shorter in length than a boat make little difference; those an inch or a foot or even a couple of feet long do no damage. Much longer waves can equally well be ignored. A surge measuring a thousand feet from crest to crest merely lifts the boat gradually up and settles it back down. But waves roughly as long as the boat are potentially destructive. This simple fact—that of the myriad of interdependent waves, only those of roughly boat length are directly threatening—lies close to the heart of physics.

Take the string theorists. These are a group of physicists with the wildly ambitious goal of bringing together the various matter-binding forces (weak, electromagnetic, strong) with gravity. Using (and occasionally inventing) new mathematics, string theory has fastened on the unimaginably small length scale of 10^{-33} cm. For at that distance, it is reckoned, the various forces no longer will be distinguishable. Instead of our familiar notion of particles—entities that resemble tiny BBs—the idea is that, looked at up close, even quarks and electrons exhibit the structure of tiny loops of string. These strings can vibrate, like plucked violin strings, at different harmonics. Each of these different "notes" corresponds to a different energy. We have known since Einstein that mass is related to energy ($E = mc^2$), so a single string could, depending on what note was playing, correspond to different masses. So perhaps, the string theorists suggest, the variety of "elementary" particles might turn out to be just different vibratory states of a single string.

At length scales much longer than 10^{-33} cm, physicists do well by ignoring the ultrasmall-scale phenomena of these strings; much of elementary particle physics is concerned with the order of the world revealed at about 10^{-16} cm. Small as that may seem, it is nearly a billion billion times larger than the world depicted by string theory. At this scale there are the observable objects, the so-called elementary particles, that leave tracks in cloud chambers, bubble chambers, or the vast electronic detectors that populate Fermilab, SLAC, and the European Organization for Nuclear Research (CERN). These particles include not only the electron, the familiar particle that governs the properties of atoms, but also heavier versions of the electron. Quarks, too, figure among the elementary particles, as do the force-carrying particles that hold matter together. And none have recently captured more interest than the elusive neutrinos, created in accelerators, our atmosphere, and stars. Through a tightly woven collaboration between experimenters and theorists, elementary-particle physicists assembled a "standard picture" of these basic particles that explained and predicted a wealth of observed effects and entities. Left out of this synthesis were certain basic questions that stubbornly haunted the community: Why do the elementary building blocks of matter—the elementary particles—have the masses they do? Why do their forces have the strengths they do, and how are they related? These are some of the fundamental questions that will be addressed in the new decade. Their ultimate resolution will involve connections to the physics of much smaller distance scales, possibly even those at play in string theory.

Moving to lengths ten thousand times larger, in the region of 10^{-12} cm, we are in the midst of nuclear physics. This is the scale at which the binding together of protons and neutrons, the hard core of an atom, is salient. Nuclear physics deals with the dynamics of the core itself, how quantum mechanics can be applied to it, how its parts undertake certain collective motions up to and including fission. It lies at the base of our understanding of fission and fusion processes, which is central to the construction of nuclear arsenals and the nuclear power industry, as well as our attempts to explain the evolution of our Sun. More recently, the Jefferson Laboratory in Newport News, Virginia, has begun investigating the nucleus in a new way, seeking to elucidate its structure in terms of the quarks in the protons and neutrons. And the Relativistic Heavy Ion Collider at Brookhaven National Laboratory, another large new facility, has begun to explore the interiors of nuclei under conditions of vastly higher energies and pressures, conditions that offer first steps toward an experimental replication of the early universe or the interior of neutron stars. But nuclear physics has more familiar and more practical consequences as well. Using electromagnetic fields to flip nuclei led to nuclear magnetic resonance, a technique that definitively sorts different substances one from the other. It has become not only a key instrument for chemistry and biochemistry but also, under the name MRI, perhaps the single most powerful medical diagnostic advance of the last half-century.

Magnetic resonance is a good jumping off point to the larger distance scale of atoms themselves. At 10^{-8} cm, atoms are ten thousand times the size of their nuclear cores. It is here that the phenomena of everyday life begin to enter. The ductility of metals, the transparency of glass, the conduction of electricity, the physical properties of matter depend crucially on the ways in which electrons move among the atoms. And it is here, in this intermediate scale between the very small and the very large, that physics has recently had some of its most astonishing advances. Over the last decade or two, atomic physicists, molecular physicists, condensed-matter physicists, and optical physicists have developed an array of techniques for observing and controlling atoms. Suddenly it becomes imaginable to design circuits by arranging the matter atom by atom. Using lasers, physicists can stretch an individual strand of DNA and other macromolecules to examine their physical properties. New materials like fullerenes, high-temperature superconductors, and magnetic materials for computer memory have tumbled from the laboratory, presenting both abstract questions about the underlying physics and an array of industrial and practical challenges. Will a deeper understanding of quantum effects like the Bose-Einstein condensate lead, for example, to a new generation of superfast quantum computers?

At the human scale, the physical world offers up new phenomena. Vortices shape the turbulence and smooth flow of both air around airplane wings and high-temperature plasmas in fusion reactors. Cracks splinter the Earth in quakes, tornadoes swirl down from the clouds, and magnetic levitation hoists trains off tracks—all at the scale of the world we measure in feet and miles. But being able to grasp something with our senses is only the first step in understanding, and a host of new physics techniques from computer simulations to chaos theory have begun to spread light on corners of nature still shaded in darkness only a generation ago.

Astrophysics and gravitational physics have in their sights the largest scale of all. For in this domain researchers are seeking order at the size of planets and stars: What are their dynamics, how do they evolve, how are chemical elements produced, how do stars exchange mass-energy with the surrounding interstellar media? Astrophysicists continue into the domain of cosmic sizes beyond planets and stars as they confront the zoo of novel phenomena discovered over the last decades: gas disks, cosmic jets, pulsars, quasars, black holes, gamma-ray bursts. Explaining these phenomena draws on physics from across the specialties, and studying them observationally has demanded the full-scale collaboration of ground-based and space-based stations. How do galaxies form? Why do they group into clusters, and clusters of clusters? What mechanism in deep space accelerates cosmic rays? What are the origin and the fate of the universe as a whole?

At all distance scales, physicists are attacking challenging problems, and as they deploy new instruments, novel concepts and deeper puzzles emerge. The physical sciences have entered a period of tremendous excitement. No one area dominates the whole, as astronomers, optical physicists, and string theorists all grapple with a vast new array of unfamiliar objects to study and an altered landscape of collaboration, along with shared instruments and techniques. Among the new collaborations is that between physics and biotechnology.

It is as evident as the daily headlines that biotechnology has altered both the medical and economic landscape. From the first days of the cyclotron, medicine and physics were bound together. Berkeley's Ernest Lawrence not only developed the earliest cyclotrons, but also used them right away to produce radioisotopes for medical purposes. And while often forgotten, Lawrence and his brother used one of their early machines to treat, successfully, their mother for a brain tumor. Since those days, the use of waves and particles to image the body and to treat tumors has become one of the essential features of modern medicine. Indeed, now that advanced computer technologies have made it possible to reconstruct electronic images of scattered waves and particles, we can see in a wide variety of ways. Scattered sound waves make ultrasound images, key not only for obstetrics but also for the examination of tumors and the study of the beating heart. Images of blood flow reveal blockages in the heart and are now providing a wealth of neurological information. CAT scans—x-ray-generated, three-dimensional cross sections of the body—go far beyond the sharpest pictures produced by traditional radiology. And once the tumor is detected, physicians now have at their disposal the full panoply of particle acceleration techniques developed over several generations. They can, depending on the location and type of pathology, irradiate the threatening growth with electrons, x rays, neutrons, protons, mesons, and heavy nuclei. Together these various techniques have propelled a vast physical-medical industry.

But the alliance between physics and the biomedical sciences goes deeper still, beyond scattering to see and scattering to kill tumors. Over the last 20 years, physicists have come to see biological matter as territory that can be explored not only in its larger structures but also atom by atom. What, we can now ask, is the precise way in which an antigen binds to an antibody? How can physics methods be used to address the complex ways in which proteins fold, and in particular how do the large biological molecules like DNA take the particular form that they do? With a precision unimaginable a generation ago, it is now routine for physicists and biologists to work together in sorting out dynamical biological processes: Individual clusters of molecules can be marked, tracked, excised, altered, inserted, or moved. Soon to come are new uses for these biomaterials in molecular motors, DNA computers, and biological elastics produced in macroscopic quantities.

Reflecting on the new relation between physics and biology, Harold Varmus, the former director of the National Institutes of Health (NIH), summed up some of the key directions for future research: an improved use of micro-manipulative methods like optical tweezers, a new and vastly more sophisticated form of data analysis closer to the work astrophysicists undertake in their deep space searches than to traditional biological research, and, finally, a biophysical attack on the signaling pathways by which cells tell each other how to respond—a task that will draw on the experience physicists have with the feedback mechanisms of complex machines. Concluding his remarks on the widening alliance between physics and biology, Dr. Varmus added:

> The NIH can wage an effective war on disease only if we—as a nation and a scientific community, not just as a single agency—harness the energies of many disciplines, not just biology and medicine. These allied disciplines range from mathematics, engineering, and computer sciences to sociology, anthropology, and behavioral sciences. But the weight of historical evidence and the prospects for the future place physics and chemistry most prominently among them.[1]

---

[1]Harold Varmus, plenary talk at the centennial meeting of the American Physical Society, Atlanta, Ga., March 22, 1999.

Modeling, imaging, miniaturizing, and controlling complex systems: These are the themes that draw together "pure" and "applied" physics in an ever-tightening weave. In national defense, many of these same themes recur. In the absence of nuclear testing, modeling the realistic characteristics of nuclear weapons is key to reliability of the arsenal. In addition to stockpile stewardship, the reduction of the global nuclear danger involves nonproliferation and arms control and the restoration of environments damaged by the production and testing of nuclear weapons. Beyond the nuclear domain, there remain vital issues of national defense, issues that if anything have gained importance in the post-Cold War epoch: cryptography, remote sensing, precision warfare, missile defense, and the development of new materials. And these are just a few of the principal areas of current research and development.

Take one arena where the interests of physicists are congruent with military importance: the Global Positioning System (GPS). Used widely in the Gulf War for directing precision weapons to their targets, the 24-satellite GPS system is based on the atomic clock. GPS had spawned a vast industry, estimated at some $2.3 billion by 1995, and generated an estimated 100,000 jobs by the end of the millennium. In a dynamic typical of physics, the GPS incorporates corrections predicted by general relativity, and military applications have diffused into the wider economy. Pilots, sailors, hikers, drivers, and surveyors all make use of what has become a device costing no more than a decent radio.

The economic fruits of physics research are visible all around us. Quantum mechanics shaped our understanding of the transistor and the electronics that pervade our world. New forms of wireless and optical technologies are shifting the way we compute, communicate, and store information. Not only is the "old" computer revolution of the postwar period now built into the infrastructure of the economy, but new waves of development continue to break, while others are just beginning to rise: biologically based computation, optical switching, and even single-molecule devices.

It is the constant exchange between understanding and application, between civilian and military, between university- and industry-based research that marks physics at the beginning of the 21st century. To prepare for the coming years in which this constant realignment of physics will no doubt continue, our older expectations of education, funding, and international cooperation all need to be reassessed. We will need a physics education and curriculum that aggressively reflect the manifold links between physics and the wider world. We will need a funding structure that captures the links between topics studied in the university, at national laboratories, and in industrial settings. And we must create modes of international cooperation that will make rational and cost-effective use of facilities too large and complex for any single nation to construct and operate.

Part I

Physics Frontiers

1

Quantum Manipulation
and New Materials

The world of quantum mechanics, although strange and beautiful, seems remote from daily experience. Yet the physicist sees quantum mechanics in action all around us, in everything from the hardness of diamond to the colors of the rainbow. The central role that quantum mechanics plays in modern technology—from the lasers that read music from compact disks to the Global Positioning System (GPS) that guides aircraft through our crowded skies—remains largely unknown to users of these devices.

We are now in the early stages of a second "quantum revolution" in which we can see and control tiny clusters of atoms and indeed even individual atoms. This second revolution is bringing together two central strands of the physics endeavor: atomic and condensed matter physics. Atomic physicists have begun to study and control the new quantum properties that emerge in large collections of atoms, a traditional theme in condensed matter physics. At the same time, condensed matter physicists have begun to learn how to shrink the materials they study to sizes in which discrete quantum excitations (the traditional province of atomic physics) play a central role. Their common meeting ground at the intersection of the microscopic quantum and the macroscopic classical worlds is rich in new physics and new technologies.

The ability to observe individual atoms and see how they assemble to form larger structures has been made possible by the development of a host of new observational tools. Scanning atomic probe microscopes can place single atoms on surfaces and measure many different physical properties of these atoms. Refinements in electron microscopy have attained single-atom resolution in bulk materials, and major advances have been achieved in the brilliance and coherence of the x-ray and neutron sources used to probe the structures of solids and large biomolecules. A new ability to control charged and neutral atoms has been made possible by the development of tools that use magnetic and laser fields to manipulate the positions and velocities of atoms in ways that are impossible using material containers. Tools like these have led to the discovery of new quantum states of matter—Bose-Einstein condensates in atomic vapors and fractional quantum Hall states in two-dimensional layers of electrons. They have shepherded in an era of control of spatially extended quantum states that shows promise for extremely precise clocks and raises the hope of using the strange properties of quantum information for radically new forms of cryptography and computation. These tools and the understanding they provide have moved us much further toward the ultimate goal of "designer" structures: objects tailored to have desired optical, mechanical, magnetic, electronic, chemical, and thermal properties.

It is astounding how far technology has progressed since the discovery of the electron only a hundred years ago. Electrons inside tiny transistors now tell the bank how much salary should be deposited in our accounts each month, and those coming along wires into our houses bring us pictures and information almost instantaneously from anywhere in the world. The ability to manipulate individual atoms will lead in the years ahead to a new generation of electronic devices that may be just as revolutionary.

NEW TOOLS FOR OBSERVATION IN THE QUANTUM REGIME

The idea of atoms is 2000 years old, yet only in the last few decades have individual atoms been seen. The optical microscope approached the limit of its resolution nearly one hundred years ago with highly polished brass and glass machines capable of seeing individual cells about 0.0001 cm in size. The resolution of such microscopes was limited by the actual size of a ray of light, its wavelength. But the revolution in quantum mechanics of the early 20th century began a new era in microscopes: Particles such as electrons can also act like waves but can have far smaller wavelengths than visible light and hence can be used to make more powerful microscopes. Modern electron microscopes have resolution 10,000 times finer than conventional optical microscopes and can see the individual atoms inside a diamond crystal.

The last 20 years have also seen the development of new types of high-resolution microscopes that scan surfaces. These instruments, which come in many varieties, use a sharp tip that is scanned across a tiny area of surface. They sense the surface much as a blind person does when reading a page of Braille text and display an image at an atomic scale. This image can be generated in response to some surface property due to, say, the force on the probe exerted by the local surface electron density. Scanning tunneling microscopes, for instance, can examine the electrons at each atom on a surface (see sidebar "Scanning Tunneling Microscope"), magnetic force microscopes can show the magnetic bits on a hard disk, near-field scanning optical microscopes probe optical properties, and atom probes can identify

SCANNING TUNNELING MICROSCOPE

The scanning tunneling microscope (STM) uses an atomically sharp metal tip to scan a surface and image the location of individual atoms. It can also be used to pick up and place single atoms in desired positions. The figure shows a "quantum corral" consisting of an elliptical ring of 34 cobalt atoms placed on a copper surface. Quantum electron waves traveling along the copper surface are reflected by the corral atoms and trapped inside. The elliptical shape was used because waves emitted from one focal point are reflected and concentrated at the other. In a remarkable recent experiment, a magnetic atom was placed at one focus so that its magnetic properties were "projected" to the other focus. The resulting "phantom atom," also known as a "quantum mirage," was detected by scanning the STM tip over the other focus. Electrons tunneling from the tip to the surface were used to probe the magnetic structure of the atom, which was faithfully reproduced at the second focus, as illustrated in the top portion of the figure. The large peak at the left is associated with the electronic structure of the atom, and the smaller peak at the right is the mirage.

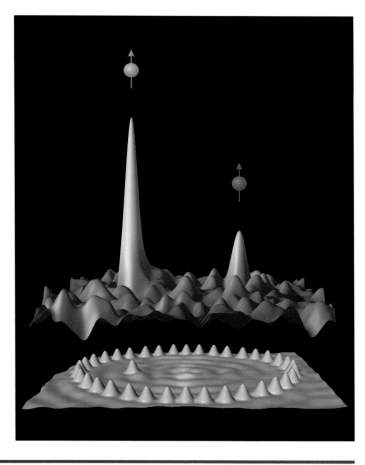

individual atoms one by one as they are pulled from a surface. Some tips can also move individual atoms around on surfaces to engineer the smallest man-made structures.

To investigate the deeper structure of materials, and in particular to examine crystal structures, x rays and neutrons can be used as probes. New high-intensity synchrotron x-ray sources have revolutionized the study of material structure. These developments in physics have important spin-offs in medicine and biology, where protein and drug structures can be analyzed routinely and quickly without the laborious work of preparing large crystals. What took years in the days when Crick and Watson were deciphering the structure of DNA now takes but a few days. Better neutron and x-ray sources that will allow much better and faster imaging are on the horizon. Spallation neutron sources, which will give information much more rapidly than current neutron sources, are coming on line, and the brightness of synchrotron x-ray sources continues to grow with time (see sidebar "Neutrons As Probes").

MANIPULATING ATOMS AND ELECTRONS

Atom Cooling and Trapping

Electric, magnetic, and laser fields are now used to confine (trap) and cool electrons and atoms, making it possible to reach far lower temperatures than ever before, as low as a billionth of a degree above absolute zero. Atoms cooled to such low temperatures change their behavior dramatically. It becomes highly nonintuitive as quantum physics dominates and new states of matter are formed: ion liquids, ion crystals, and atomic Bose-Einstein condensates. Strange new quantum states can be created, providing arenas in which to test basic quantum mechanics and novel methods of quantum information processing.

The most widespread advances have come from using laser radiation to slow, and hence cool, a sample of atoms. Radiation pressure arises from the kick an atom feels when light scatters off it. If a sample of atoms is irradiated with beams from all six directions and the color of the light is carefully adjusted to the proper value, one creates "optical molasses." Because of the Doppler shift, all the atoms in this molasses preferentially scatter photons that are opposing their motion, and the atoms are quickly slowed. This results in temperatures of less than one thousandth of a degree above absolute zero. A variety of methods have been found for cooling isolated atoms to even lower temperatures. In so-called Sisyphus laser cooling, precooled

NEUTRONS AS PROBES

The neutron is an electrically neutral particle existing in the nucleus of atoms along with its positively charged counterpart, the proton. Because it is electrically neutral, the neutron can penetrate into the nucleus of atoms and so is an especially useful experimental probe of the structure of materials. It is very sensitive to hydrogen atoms and can be used to locate them precisely within complex molecules, enabling a more accurate determination of molecular structure, which is important for the design of new therapeutic drugs. Because large biological molecules contain many hydrogen atoms, the best way to see part of a biomolecule is through isotope substitution—replacing hydrogen with heavy hydrogen (deuterium) atoms. Deuterium atoms and hydrogen atoms scatter neutrons differently. Thus, in a technique called contrast variation, scientists can highlight different types of molecules, such as a nucleic acid or a protein in a chromosome, and glean independent information on each component within a large biomolecule. We see below a model of two insulin molecules containing zinc ions (white balls). Insulin molecules pick up zinc ions when crystallized for neutron diffraction studies.

Neutrons for use in basic and applied research are produced in nuclear reactors or with particle accelerators. Most of the world's neutron sources were built decades ago, and although the uses and demand for neutrons have increased throughout the years, few new sources have been built. But now the U.S. Congress, through the Department of Energy's Office of Science, has funded the construction of a new, accelerator-based neutron source, the Spallation Neutron Source (SNS) (below) at Oak Ridge National Laboratory, which will provide the most intense pulsed neutron beams in the world for scientific research and industrial development. When complete in 2006, the SNS will be over 10 times more powerful than the best existing pulsed neutron source in the world. Its innovative set of instruments will enable forefront research that will benefit industry and health and increase our understanding of the underlying structure of materials.

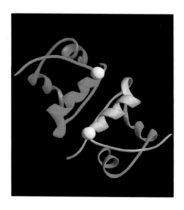
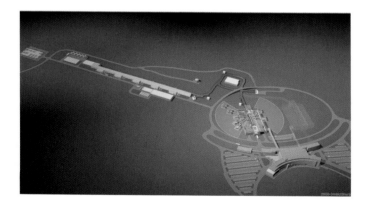

atoms moving through intersecting laser beams feel as though they are climbing hills, causing them to slow down no matter in which direction they are moving. This process can cool atoms to a few millionths of a degree above absolute zero. The very lowest temperatures (billionths of a degree) are achieved by the nonlaser technique of evaporative cooling of trapped atoms. This is analogous to the way a cup of hot coffee cools, by giving off the most energetic molecules as steam. As the energetic atoms escape from the trap (or coffee cup) the remaining atoms become very cold.

As in any refrigerator, to cool an object, whether it be a cloud of atoms or a carton of milk, the object must be insulated from the hotter outside world. Traps based on laser and magnetic fields can now do this for small numbers of isolated atoms in a much more controlled and gentle way than is possible in a normal material container. Magnetic bottles use properly shaped magnetic fields to hold atoms with large magnetic moments. In the dipole force laser trap, atoms are sucked into the center of a focused laser beam. This laser trap also works on much larger objects, such as living cells or large molecules. These optical tweezers, as they are now called, have moved from the atomic physics lab to the biology lab, where they are used routinely to manipulate cells and large biological molecules such as DNA.

The workhorse of ultracold neutral atom research is the magneto-optical laser trap, attractive because of its simplicity of construction and its ability to simultaneously cool and confine. The radiation pressure from laser beams converging on a center creates the trap. A weak magnetic field is used in a subtle way to control this pressure such that the atoms are all pushed toward the trap center and cooled. The laser trap provides a simple and inexpensive source of very cold trapped atoms and is seeing widespread use in the research laboratory as well as in novel applications for improving atomic clocks and lithography.

Achieving and Using the Quantum Regime

Novel cooling and confinement techniques allow atoms to be sufficiently cooled so that they no longer act as a group of indistinguishable particles but rather as waves corresponding to a single coherent quantum state. This produces simple quantum systems that have much larger spatial extents (macroscopic) than were possible in the past and atomic motion that is entirely wavelike. The superior insulation of electromagnetic atom traps also means that the electronic energy levels of the atom are perturbed much less than in conventional containers. This makes it possible to create and preserve particular, desired quantum states and to measure atomic and electronic properties much more precisely.

Experiments on individual trapped electrons have measured their magnetic moment, which is due to their spin, to an accuracy of better than 5 parts in 10^{12} (1000 billion). The value of the magnetic moment obtained from these measurements agrees with that predicted by the fundamental theory of electrons and light—quantum electrodynamics. This successful comparison of experimental data with theoretical calculations represents the most stringent test of a theory in all of science. Atomic structure and interactions have also been measured much more accurately than before using such cooled and trapped samples. A special case has been the atomic clock, a device using the structure of the atom to define the second precisely (see sidebar "Atomic Clocks"). The world's most precise clocks now use cesium fountains, in which samples of ultracold cesium atoms are launched upward. As the atoms ascend and then fall back to the source, the clock transition is measured with an inaccuracy of only 2 parts in 10^{15} (7 millionths of a second in a century).

The cooling of atoms to temperatures where they behave like quantum mechanical waves has spurred the development of atomic counterparts to optical lenses, mirrors, and diffraction gratings, allowing atomic beams to be reflected, focused, split, and recombined in much the same way as light beams. In these devices it is the quantum waves of the atoms, rather than light waves, that are interfering. Optical interferometer gyroscopes are now widely used in navigational systems, but their atom interferometer counterparts can in principle be far more sensitive. Atom-wave interferometers also have measurement capabilities for which there is no optical analogue, such as the sensitive detection of electric, magnetic, and gravitational fields.

Gaseous Bose-Einstein Condensates

An exciting outcome of atom cooling and trapping techniques has been the creation of a novel form of matter, the gaseous Bose-Einstein condensate (BEC). Einstein predicted this effect in 1925, and superfluid helium and superconductivity are manifestations of it in liquids and solids. However, the original concept of Einstein, the possibility of condensation in a dilute atomic gas, was not realizable at that time, requiring temperatures much colder than could be attained. BEC in a gas was finally realized in 1995 by cooling a cloud of rubidium atoms to less than 100 billionths of a degree above absolute zero, and it has been widely duplicated since then using a

ATOMIC CLOCKS

The atomic clock is a wonderful example of the unanticipated benefits from basic physics research. Originally physicists were interested in measuring the separation of atomic energy levels to a higher precision to understand better the physics of atoms. They soon found that their experiments were more accurate than the clocks used to determine the frequency of the microwaves that were used to excite the atoms. This realization led to the atoms themselves becoming the time standard. The second is now defined in terms of the separation of the two lowest energy states in the cesium atom.

Atomic clocks provide the basis for precise navigation, including the global positioning system. The GPS is based on a set of orbiting satellites, each carrying a very precise atomic clock. A GPS receiver uses the exact time that it receives a clock signal from each satellite to tell the distance from the satellite. By knowing the distance to several satellites, the position of the receiver can be determined. Although originally implemented for its military uses, the GPS is now in wide use for nonmilitary applications. Hand-held units costing a few hundred dollars can give a precise location anywhere on Earth to less than a hundred feet.

Atomic cesium fountain clock, accurate to 1 second in 20 million years.

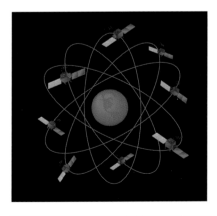

24 GPS satellites orbit the Earth.

variety of different atoms. It was accomplished using the techniques of laser cooling and trapping combined with magnetic trapping and evaporative cooling. At these low temperatures, a large fraction of the atoms goes into the lowest-energy quantum state allowed in the atom trap.

BEC in a gas is proving to be a fascinating new system because it is a quantum state with very special properties. It is enormously large (about the diameter of a human hair and containing millions of atoms) compared with quantum states in other systems, so it can be observed and manipulated in a way that had not been possible until now. Adding visible light and/or microwaves to the magnetic traps is proving to be a particularly convenient way to manipulate the condensates. A number of such techniques have been found to eject the condensate samples from the trap to obtain bright atomic beams, so-called "atom lasers." Finally, the behavior of the condensate is amenable to theoretical analysis because the interactions between the atoms involved are very well understood.

BEC has opened up opportunities to explore quantum behavior in a novel regime as well as allowed manipulation of atomic samples with a precision limited only by the uncertainty principle of quantum mechanics. Experiments on condensate behavior involving dissipation and coupling between the quantum state and the environment have provided many surprises and challenges to theoretical understanding. This work will lead to a much better understanding of dissipation in quantum mechanics, and it connects to the studies on the fragility of quantum entanglement that are of paramount importance for the future of quantum computing and quantum-limited measurement instruments.

Laser Control of Electronic States

The quantum engineering of BEC involves primarily the motion of atoms as a whole and not of their electrons, because it is difficult to compete with the binding forces within the atom that control the electron. However, this is beginning to change as a result of recent improvements in the laser technology used in producing extremely intense short pulses of light. Table-top systems can produce pulses of light so intense that the electric fields of the light can not only be as strong as those found inside atoms and molecules, but can also be shorter than the time scale for the motion of the electrons inside the atoms and molecules.

An atom or molecule in such a light field is really no longer an atom or a collection of atoms but rather a new regime of matter, with the electrons, atomic nuclei, and light field having equal roles in determining the structure and behavior. There is still much to be understood about this system. To control the electron dynamics and consequently the interatomic interactions requires the detailed manipulation of the time dependence of the light pulses. Clever techniques for controlling and observing the behavior of light pulses on time scales of 10^{-15} seconds are being developed to make this possible. This control is still very limited, and much of the basic physics of these systems is yet to be understood. The ultimate goal will be to use laser light to control the chemical processes at the atomic scale—to enable chemists to, say, produce efficiently a desired molecule or a new chemical compound.

Another practical use of the novel behavior of atoms in intense laser fields has been the creation of brighter sources of light in the far-infrared, ultraviolet, and x-ray spectral regions with shorter pulses. Pulsed x rays produced in this interaction between atoms and intense light will make it possible to study problems in surface science, such as catalysis, chemistry, atomic physics, and biological imaging, with unprecedented temporal and spatial resolution. Techniques using these infrared pulses are being developed for monitoring the water content of leaves and food, for distance-ranging applications, and for medicine. The use of these novel laser sources in future particle accelerators is also being pursued.

NEW MATERIALS

The ability to create new materials and structures is inextricably linked to advances in the understanding of fundamental phenomena in materials physics. These advances, along with improvements in synthesis and processing, have led to an astonishing array of new materials with unexpected properties and to dramatic improvements in the properties of established materials. Some of these developments have provided fertile ground for the exploration of novel fundamental phenomena; others show promise for finding applications quickly. Some even have the potential to change our lives. Entirely new and unexpected phenomena often appear in a new material. Layered cuprate high-temperature superconductors are a new class of materials that has kept experimentalists and theorists searching to understand the physical basis of high-temperature superconductivity (see sidebar "High-Temperature Superconductivity"). This basis appears to be very different from that of conventional superconductors. New materials allow entirely new device concepts to be realized or lead to a dramatic change in scale, such as single-molecule wires made of carbon nanotubes. Semiconductor nanoclusters, which emit light whose wavelength depends on cluster

HIGH-TEMPERATURE SUPERCONDUCTIVITY

High-temperature superconductors, first discovered in the 1980s, are materials that can conduct electrical current without resistance at temperatures far above those possible with any superconductors known up to that time. This group of materials, known as cuprates, contain planes that are networks of copper and oxygen as well as other elements that control the number of electrons in each copper-oxygen plane. Understanding the behavior of cuprates continues to be one of the most formidable challenges in modern science. Ever more sophisticated experimental probes and ever more refined materials preparation yield overwhelming evidence that novel concepts are needed to describe these materials. Cuprate superconductors have unusual symmetry properties. Electrons in all superconductors form pairs, and in most superconductors the distribution of electrons in the pair is completely symmetric. In cuprates, however, the electron distribution is less symmetric, as illustrated by the fourfold symmetry in the figure below. Today's high-temperature superconductors are finding growing use in filters for the wireless communications industry, as magnetic field sensors in medical scanning applications, and in electrical transmission cables for high-power, high-current applications.

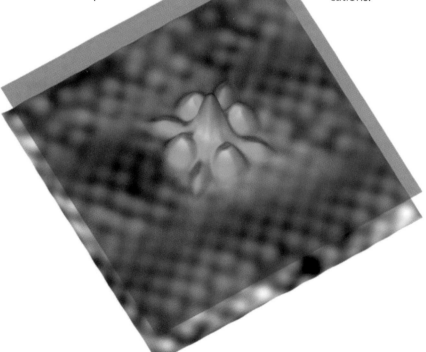

size, allow the tailoring of material properties to suit a particular need. Even mature techniques, such as those for crystal growth, demand continuous improvements in process control to produce the material required for technological applications and fundamental studies.

Many of the new materials and structures are far more complex than those that were studied previously, requiring advances in processing to allow control of the increased complexity. Some involve the synthesis of an entirely new compound or material with unexpected properties. In other cases, advances in processing have allowed the fabrication of new or modified materials or structures whose properties were suspected before the material was actually made. This may allow a well-known compound to be remade in a new form with different properties. Finally, well-known materials sometimes exhibit new (and in some cases unexpected) properties that appear when the ability to process them is improved.

Many of the materials advances listed in Table 1.1 addressed a technological need, such as information storage and transfer. Others were driven

TABLE 1.1   Some New Materials of the Past 15 Years

Advance	Driver	Nature of Advance
New compounds/materials		
High-temperature superconductors	Science	Unexpected
Organic superconductors	Science	Unexpected
Rare-earth optical amplifiers	Technology	Evolutionary
High-field magnets	Technology	Evolutionary
Organic electronic materials	Technology	Evolutionary
Magneto-optical recording materials	Technology	Evolutionary
Amorphous metals	Technology	Evolutionary
New structures of known materials		
Quasicrystals	Science	Unexpected
Buckyballs and related structures	Science	Unexpected
Nanoclusters	Science	Evolutionary
Metallic hydrogen	Science	Evolutionary
Bose-Einstein condensates	Science	Evolutionary
Giant magnetoresistance materials	Technology	Unexpected
Diamond films	Technology	Evolutionary
Quantum dots	Technology	Evolutionary
Foams/gels	Technology	Evolutionary
New properties of known materials		
Gallium nitride	Technology	Unexpected
Silicon-germanium	Technology	Evolutionary

by scientific curiosity. Many discoveries that result from pure scientific curiosity ultimately find their way into products. Low-temperature super-conductors are now used in magnets for magnetic resonance imaging. Other discoveries, though originally motivated by a technological need, give rise to very beautiful and fundamental insights. The fractional quantum Hall effect was first observed in high-mobility semiconductor structures now used in high-frequency applications. Many of the advances listed in Table 1.1, such as the development of giant magnetoresistance materials, have driven the technology of the information age. Several of these are described in Chapter 10.

ARTIFICIAL NANOSCALE STRUCTURES

While atomic physicists have learned how to hold small numbers of atoms or electrons in vacuum, solid-state physicists have begun to contain atoms and electrons in tiny volumes within solids. These structures have many practical applications. Through them we control the movement of electrons that is the basis of electronics. As the structures being explored for new electronic devices become smaller than the wavelength of light, one enters the realm of nanoscale physics, a field that is rapidly evolving and expanding as the size of structures decreases and the quality and complex-ity increase.

Fabrication

New tools for materials fabrication have now emerged that permit cut-ting and pasting almost on the atomic scale. Molecular-beam epitaxy (MBE) is the gentlest spray-painting technique known to man. With a beam of atoms streaming against a surface in vacuum, layers as thin as one atom can be placed on top of one another. The method—developed in the last few decades because of the need for very high speed transistors, such as those used to send and receive radio waves from cellular phones—has had the additional benefit of opening up new areas in basic physics. The developers of MBE could not have imagined that this tool would lead to the discovery of a fascinating new phenomenon that led to the Nobel Prize in physics in 1998—the fractional quantum Hall effect. Because of the thin layers and high perfection of the MBE-created samples, a pancake-like gas of electrons can travel almost without collisions inside a solid. When cooled in a high magnetic field, these electrons act collectively like particles whose charge is a fraction of the charge on a single electron.

The nanofabrication challenge for the future is to achieve control in all three dimensions, rivaling the layer precision of MBE. This is like controlling not just the thickness of the layer of paint but also the width of the brush, requiring it to be atom-sized. Technology here has made great strides. In a silicon chip, such as a state-of-the-art microprocessor, the smallest features are now about 200 nm, one thousandth the width of a human hair. Extensions of the tools developed to make these chips allow structures as small as 10 nm to be fabricated. The real challenge is to control these structures with even greater precision. Also, ways must be found to reduce the damage induced by the microscopic cutting tools used to make nanostructures, because the damaged surfaces can perturb the desired properties by disrupting the movement of the electrons. As the individual atomic scale is reached, a new approach—perhaps using naturally stable structures or relying on self-assembly—will be needed.

Self-assembly

Many structures in nature are well organized on the nanoscale. For example, a seashell has a complex interleaved structure with exceptional strength yet low mass. Opals are perfectly arranged microspheres of silica. Materials physicists have increasingly come to view sophisticated forms of self-assembly as unique tools to control nanoscale materials. So-called diblock copolymers are a beautiful example from chemistry: polymer blends that give perfectly organized and highly controlled structures on the nanoscale.

When atoms are deposited by MBE on some surfaces, they do not form a uniform film but instead ball up to form islands. This effect, seen for example when germanium is deposited on silicon, has only recently begun to be understood. The stress from the differences in atom size causes the germanium islands to repel one another and to self-organize. This will have very important applications—for example, in semiconductor lasers—because of the unique quantum effects when the lasing regions are confined on the nanoscale.

A challenge for self-assembly is to reach the perfection often required for specific physical properties. This is the primary reason basic research is needed to gain a better understanding of these complex phenomena. The nanoscale perfection found in biology encourages the belief that this is achievable. Biology and physics will overlap even more in the future as physicists imitate nature for improved nanostructural control and as artificial nanostructures are used in biomedical applications.

Nanoscale and Molecular Electronics

The study of nanoscale electronic devices began to blossom in the last decade and a half. It is now possible to fabricate devices that are so tiny that the charging energy needed to add or remove a single electron becomes easily observable. In some cases even the spacing of individual electron energy levels is large enough to be discernible, making these devices analogous to artificial atoms. These technological advances have allowed physicists to address such fundamental issues as how small a superconducting grain can be and still remain a superconductor. They have also permitted the construction of practical devices such as precision charge pumps that use a bucket-brigade technique to pump exactly one electron through the device for each cycle of an applied voltage. A remarkable new invention, the radio frequency single-electron transistor, uses transport through a nanoelectronic device to measure the presence of nearby tiny electrical charges with unprecedented speed and sensitivity.

A related and important new direction for the coming decade can be called "molecular electronics." Using methods made possible by advances in lithographic fabrication and scanning tunneling microscopy, physicists are beginning to study the transport of electrons through a single molecule. In the future it may be possible to identify DNA sequences from their electrical properties. A novel problem of great interest is the transport of electrons through carbon nanotubes. These "molecular wires" are essentially narrow strips of graphite rolled into a single tube a billionth of a meter in diameter, forming a highly elongated molecule of pure carbon. It is now possible to make electrical connections to these tubes and inject electrons into them. Small changes in the way the graphite strip is rolled up cause large changes in the current flow, allowing the fabrication of tubes that are metallic, insulating, or semiconducting. This suggests the possibility of ultraminiaturized transistor-like devices based on these novel molecules.

QUANTUM INFORMATION AND
THE ENGINEERING OF ENTANGLED STATES

Fundamentally, quantum mechanics is a theory of information. Quantum information is radically different from its classical counterpart, which forms the basis of the current information age. Recent progress in the understanding and manipulation of quantum information has brought together physicists from several diverse areas in a quest to advance the frontiers of communication and computation. The exciting but still speculative hope is that their work will lead to important new information technologies.

The smallest unit of classical information, the bit, consists of the answer to a single yes-no (or true-false) question. In the current digital age, bits are represented by tiny electrical switches that are either open or closed (on or off). Even the most advanced computer is nothing more than a series of such switches wired together so that the position of one switch controls the position of others. The positions of a set of input switches ultimately determine the positions of a set of output switches that represent the result of the computation.

By contrast, a quantum bit of information ("qubit") is encoded in a quantum switch whose position is not fully knowable. That is, the switch can be in a coherent superposition of the two possibilities, open and closed, and measurement of its position destroys the quantum information. From a classical information point of view, it would seem that a precise output is not possible. However, a surprising discovery of the last decade was the existence of quantum algorithms that can provide imprecise but good-enough solutions far faster than their classical counterparts. This advantage comes because the quantum computer operates on all the different possible switch positions at the same time rather than having to consider them each sequentially. Quantum algorithms have received a great deal of attention since the discovery of an algorithm to factor large numbers, the basis of much of modern encryption (see sidebar "Quantum Cryptography").

A realistic quantum computer must be based on switches built out of electrons and nuclei of individual atoms, or nanoscopic condensed matter systems. The last decade has seen great advances in the ability to manipulate such systems and to prepare the special quantum states required. These are so-called entangled quantum systems, in which two or more quite different and apparently unrelated properties of an object (usually an atom) are linked. The motion of the object might be entangled with its color, in which case measuring the motion of the object affects the color that one would subsequently measure, or vice versa. The widespread creation and study of entangled quantum states have only recently been made possible by the delicate manipulation of special systems, such as trapped atoms and tiny superconducting metal grains. Using combinations of laser light and radio waves to manipulate the trapped atoms, rudimentary computational operations have been performed on qubits.

Quantum engineering methods can be extended from a single atom to entangled states of many atoms using a suitable coupling mechanism. The strong electrical interaction between trapped charged atoms provides such a mechanism, and four-atom entangled states (four qubits) have recently been demonstrated. Physicists are still exploring the fundamental science in

QUANTUM CRYPTOGRAPHY

In the modern information age, there has been an almost complete convergence between the technologies for computation and communication. Similarly, recent advances in the understanding of quantum computation have gone hand in hand with great strides in the transmission of quantum information.

One application of quantum information that is already becoming practical is the secure transmission of encryption codes ("quantum cryptography"). The biggest obstacle to quantum computing is that any interaction with the environment disrupts the quantum information. For quantum cryptography this fragility is a virtue, because it means that it is impossible for an eavesdropper to intercept the code signal without disrupting its quantum information so much that the interception is easily apparent. Thus, quantum cryptography uses the laws of physics to ensure that there is no undetected interception of a coded message. This has now been demonstrated using entangled states of photons to securely transmit cryptographic keys over distances of several kilometers. Demonstrations in Switzerland have sent the appropriately prepared photons of light down an optical fiber, and in the United States the signals were sent through open air between two locations near Los Alamos National Laboratory.

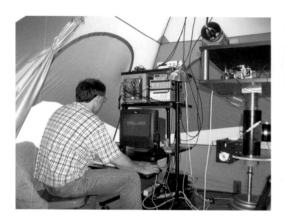 

At the left is an operator sending a signal on the free-space transmitter; at the right, the signal, consisting of photons, is being received.

these technologies to learn what may or may not be feasible for practical applications. Although practical quantum computers are not likely to exist in the near future, more modest uses for multi-atom entangled states, such as improvements in atomic spectroscopy and atomic clocks, appear to be within reach.

We are in the midst of an exciting revolution in the ability to observe and manipulate material at the quantum level. The next few decades are certain to lead to new insights into the strange world of quantum physics and to dramatic advances in technology, as the field of quantum engineering is developed.

2

Complex Systems

Life, our environment, and the materials all around us are complex systems of interacting parts. Often these parts are themselves complex. Take the human body. At a very small scale it is just electrons and nuclei in atoms. Some of these are organized into larger molecules whose interactions produce still larger-scale body functions such as growth, movement, and reasoning. The study of complex systems presents daunting challenges, but rapid progress is now being made, bringing with it enormous benefits. Who can deny, for instance, the desirability of understanding crack formation in structural materials, forecasting catastrophes such as earthquakes, or treating plaque deposition in arteries?

A traditional approach in physics when trying to understand a system made up of many interacting parts is to break the system into smaller parts and study the behavior of each. While this has worked at times, it often fails because the interactions of the parts are far more complicated and important than the behavior of an individual part. An earthquake is a good example. At the microscopic level an earthquake is the breaking of bonds between molecules in a rock as it cracks. Theoretical and computational models of earthquake dynamics have shown that earthquake behavior is characterized by the interacting of many faults (cracks), resulting in avalanches of slippage. These slippages are largely insensitive to the details of the actual molecular bonds.

In the study of complex systems, physicists are expanding the boundaries of their discipline: They are joining forces with biologists to understand life, with geologists to explore Earth and the planets, and with engineers to study crack propagation. Much progress is being made in applying the quantitative methods and modeling techniques of physics to complex systems, and instruments developed by physicists are being used to better measure the behavior of those systems. Advances in computing are responsible for much of the progress. They enable large amounts of data to be collected, stored, analyzed, and visualized, and they have made possible numerical simulations using very sophisticated models.

In this chapter, the committee describes five areas of complex system research: the nonequilibrium behavior of matter; turbulence in fluids, plasmas, and gases; high-energy-density systems; physics in biology; and Earth and its surroundings.

NONEQUILIBRIUM BEHAVIOR OF MATTER

The most successful theory of complex systems is equilibrium statistical mechanics—the description of the state that many systems reach after waiting a long enough time. About a hundred years ago, the great American theoretical physicist Josiah Willard Gibbs formulated the first general statement on statistical mechanics. Embodied in his approach was the idea that sufficiently chaotic motions at the microscopic scale make the large-scale behavior of the system at, or even near, equilibrium particularly simple. Full-scale nonequilibrium physics, by contrast, is the study of general complex systems—systems that are drastically changed as they are squeezed, heated, or otherwise moved from their state of repose. In some of these nonequilibrium systems even the notion of temperature is not useful. Although no similarly general theory of nonequilibrium systems exists, recent research has shown that classes of such systems exhibit patterns of common ("universal") behavior, much as do equilibrium systems. These new theories are again finding simplicity in complexity.

Everyday matter is made of vast numbers of atoms, and its complexity is compounded for materials that are not in thermal equilibrium. The properties of a material depend, then, on its history as well as current conditions. Although materials in thermal equilibrium can display formidable complexity, nonequilibrium systems can behave in ways fundamentally different from equilibrium ones. Nature is filled with nonequilibrium systems: Looking out a window reveals much that is not in thermal equilibrium, including all living things. The glass itself has key properties, such as transparency and strength, that are very different from those of the same material (SiO_2) in the crystalline state. Essentially all processes for manufacturing industrial materials involve nonequilibrium phenomena.

A few of the many examples of matter away from equilibrium are described below.

Granular Materials

Granular materials are large conglomerations of distinct macroscopic particles. They are everywhere in our daily lives, from cement to cat food to chalk, and are very important industrially; in the chemical industry approximately one-half of the products and at least three-quarters of the raw materials are in granular form. Despite their seeming simplicity, granular materials are poorly understood. They have some properties in common with solids, liquids, and gases, yet they can behave differently from these familiar forms of matter. They can support weight but they also flow like liquids—sand on a beach is a good example. A granular material can behave in a way reminiscent of gases in that it allows other gases to pass through it, such as in fluidized beds used industrially for catalysis. The wealth of phenomena that granular materials display cannot easily be understood in terms of the properties of the grains of which they are composed. The presence of friction combines with very-short-range intergrain forces to give rise to their distinctive properties.

Glasses: Nonequilibrium Solids

When a liquid is cooled slowly enough, it crystallizes. However, most liquids, if cooled quickly, will form a glass. The glass phase has a disordered structure superficially similar to a liquid, but unlike the liquid phase, its molecules are frozen in place. Glasses have important properties that can be quite different from those of a crystal made of the same molecules—properties, for example, that are vital to their use as optical fibers. Understanding the relationship between the properties of a glass, the atoms of which it is composed, and the means by which it was prepared is key for the development and exploitation of glasses with useful properties. Understanding the nature of the glass transition (whether it is a true phase transition or a mere slowing down) is of fundamental importance and will yield insight into other disordered systems such as neural networks.

Failure Modes: Fracture and Crack Propagation

Understanding the initiation and propagation of cracks is extremely important in contexts ranging from aerospace and architecture to earthquakes. The problem is very challenging because cracking involves the interplay of many length scales. A crack forms in response to a large-scale strain, and yet the crack initiation event itself occurs at very short distances.

In many crystals the stress needed to initiate a crack drops substantially if there is a single individual defect in the crystal lattice. This is one example of the general tendency of nonequilibrium systems to focus energy input from large scales onto small, sometimes even atomic scales. Other examples include turbulence, adhesion, and cavitation. Depending on material properties that are not well understood, a small crack can grow to a long crack, or it can blunt and stop growing. Recent advances in computation and visualization techniques are now providing important new insight into this interplay through numerical simulation.

Foams and Emulsions

Foams and emulsions are mixtures of two different phases, often with a surfactant (for instance, a layer of soap) at the boundary of the phases. A foam can be thought of as a network of bubbles: It is mostly air and contains a relatively small amount of liquid. Foams are used in many applications, from firefighting to cooking to shaving to insulating. The properties of foams are quite different from those of either air or the liquid; in fact, foams are often stiffer than any of their constituents. Emulsions are more nearly equal combinations of two immiscible phases such as oil and water. For example, mayonnaise is an emulsion. The properties of mayonnaise are quite different from those of either the oil or the water of which it is almost entirely composed. With the widespread use of foams and emulsions, it is important to design such materials so as to optimize them for particular applications. Further progress will require a better understanding of the relationship between the properties of foams and emulsions and the properties of their constituents and more knowledge of other factors that determine the behavior of these complex materials.

Colloids

Colloids are formed when small particles are suspended in a liquid. The particles do not dissolve but are small enough for thermal motions to keep them suspended. Colloids are common and important in our lives: paint, milk, and ink are just three examples. In addition, colloids are pervasive in biological systems; blood, for instance, is a colloid.

A colloid combines the ease of using a liquid with the special properties of the solid suspended in it. Although it is often desirable to have a large number of particles in the suspension, particles in highly concentrated suspensions have a tendency to agglomerate and settle out of the suspension.

The interactions between the particles are complex because the motions of the particles and the fluid are coupled. When the colloidal particles themselves are very small, quantum effects become crucial in determining their properties, as in quantum dots or nanocrystals.

New x-ray- and neutron-scattering facilities and techniques are enabling the study of the relation between structure and properties of very complex systems on very small scales. Sophisticated light-scattering techniques enable new insight into the dynamics of flow and fluctuations at micron distances. Improved computational capabilities enable not just theoretical modeling of unprecedented faithfulness, complexity, and size, but also the analysis of images made by video microscopy, yielding new insight into the motion and interactions of colloidal particles.

The combination of advanced experimental probes, powerful computational resources, and new theoretical concepts is leading to a new level of understanding of nonequilibrium materials. This progress will continue and is enabling the development and design of new materials with exotic and useful properties.

TURBULENCE IN FLUIDS, PLASMAS, AND GASES

The intricate swirls of the flow of a stream or the bubbling surface of a pan of boiling water are familiar examples of turbulence, a challenging problem in modern physics. Despite their familiarity, many aspects of these phenomena remain unexplained. Turbulence is not limited to such everyday examples; it is present in many terrestrial and extraterrestrial systems, from the disks of matter spiraling into black holes to flow around turbine components and heart valves. Turbulence can occur in all kinds of matter, for example, molten rock, liquids, plasmas, and gases.

The understanding and control of turbulence can also have significant economic impact. For instance, the elimination or reduction of turbulence in flow around cars reduces fuel consumption, and the elimination of turbulence from fusion plasmas reduces the cost of future reactors (see sidebar "Suppressing Turbulence to Improve Fusion"). Perhaps the most dramatic phenomena involving turbulence are those with catastrophic dynamics, such as tornadoes in the atmosphere, substorms in the magnetosphere, and solar flares in the solar corona.

Although turbulence is everywhere in nature, it is hard to quantify and predict. There are four important reasons for this. First, turbulence is a complex interplay between order and chaos. For example, coherent eddies in a turbulent stream often persist for surprisingly long times. Second, turbu-

SUPPRESSING TURBULENCE TO IMPROVE FUSION

To produce a practical fusion energy source on Earth, plasmas with temperatures hotter than the core of the Sun must be confined efficiently in a magnetic trap. In the 1990s, new methods of plugging the turbulent holes in these magnetic cages were discovered. The figure below shows two computer simulations. The leftmost simulation shows turbulent pressure fluctuations in a cross section of a tokamak—a popular type of magnetic trap. These turbulent eddies are like small whirlpools, carrying high-temperature plasma from the center toward the lower-temperature edge. The rightmost simulation shows that adding sheared rotational flows can suppress the turbulence by stretching and shearing apart the eddies, analogous to keeping honey on a fork by twirling the fork.

Several experimental fusion devices have demonstrated turbulence suppression by sheared rotational flows. This graph of data from the Tokamak Fusion Test Reactor at the Princeton Plasma Physics Laboratory, which achieved 10 MW of fusion power in 1994, shows how suppressing the turbulence can significantly boost the central plasma pressure and thus the fusion reaction rate.

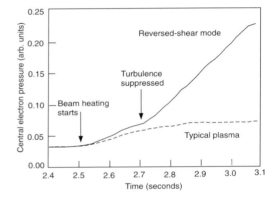

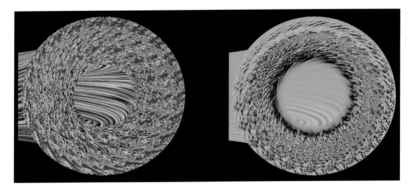

lence often takes place over a vast range of distances. Third, the traditional mathematical techniques of physics are inappropriate for the highly nonlinear equations of turbulence. And fourth, turbulence is hard to measure experimentally.

Despite these formidable difficulties, research over the past two decades has made much progress on all four fronts: For example, ordered structures have been identified within chaotic turbulent systems. Such structures are often very specific to a given system. Turbulence in the solar corona, for example, appears to form extremely thin sheets of electric current, which may heat the solar corona. In fusion devices, plasma turbulence has been found theoretically and experimentally to generate large-scale sheared flows, which suppress turbulence and decrease thermal losses from the plasma. The patchy, intermittent nature of turbulence so evident in flowing streams is being understood in terms of fractal models that quantify patchiness. The universe itself is a turbulent medium; much of modern cosmological simulation seeks to understand how the observed structure arises from fluctuations in the early universe.

The recognition that turbulent activity takes place over a range of vastly different length scales is ancient. Leonardo da Vinci's drawings of turbulence in a stream show small eddies within big ones. In theories of fluid turbulence, the passing of energy from large to small scales explains how small scales arise when turbulence is excited by large-scale stirring, like a teaspoon stirring coffee in a cup. Small-scale activity can also arise from the formation of isolated singular structures: shock waves, vortex sheets, or current sheets.

The difficulty of measuring turbulence is due partly to the enormous range of length scales and the need to make measurements without disturbing the system. However, the structure and dynamics of many turbulent systems have been measured with greatly improved accuracy using new diagnostic tools. Light scattering is being used to probe turbulence in gases and liquids. Plasma turbulence is being measured by scattering microwaves off the turbulence and by measuring light emission from probing particles. Soft x-ray imaging of the Sun by satellites is revealing detailed structure in the corona. Measurements of the seismic oscillations of the Sun are probing its turbulent convection zone. Advances in computing hardware and software have made it possible to acquire, process, and visualize vast turbulence data sets (see sidebar "Earth's Dynamo").

HIGH-ENERGY-DENSITY SYSTEMS

When stars explode, a burst of light suddenly appears in the sky. In a matter of minutes, the star has collapsed and released huge quantities of energy. These supernova explosions are examples of high-energy-density phenomena, an immense quantity of energy confined to a small space. Our

EARTH'S DYNAMO

Earth's magnetic field reverses direction on average every 200,000 years. The series of snapshots of a reversal shown below is taken from a computer simulation in which such reversals occur spontaneously. The snapshots are 300 years apart during one of two reversals in the 300,000 years simulated. In the top four pictures the red denotes outward field and the blue inward field on Earth's surface. The bottom four show the same thing beneath the crust on top of the liquid core. Through computer simulation this complex phenomenon is being understood and quantified.

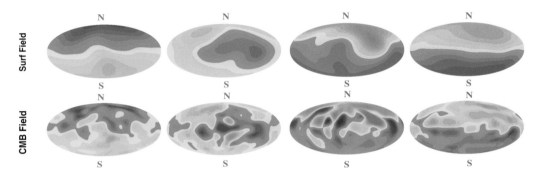

Motions of Earth's liquid core drive electrical currents that form the magnetic field. This field pokes through Earth's surface and orients compasses. The picture at the right shows the magnetic field lines in a computer simulation of this process. Where the line is blue it is pointing inward and where it is gold it is pointing outward. The picture is two Earth diameters across.

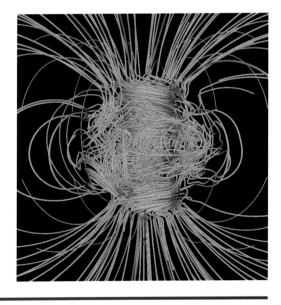

experience with high energy densities has been limited; until recently, it was mainly the concern of astrophysicists and nuclear weapons experts. However, the advent of high-intensity lasers, pulsed power sources, and intense particle beams has opened up this area of physics to a larger community.

There are several major areas of research in high-energy-density physics. First, high-energy-density regimes relevant to astrophysics and cosmology are being reproduced and studied in the laboratory. Second, researchers are inventing new sources of x rays, gamma rays, and high-energy particles for applications in industry and medicine. Third, an understanding of high-energy-density physics may provide a route to economical fusion energy. And it is worth noting that the very highest energy densities in the universe since 1 μs after the Big Bang are being created in the Relativistic Heavy Ion Collider (RHIC) at Brookhaven National Laboratory.

Since 1980, the intensity of lasers has been increased by a factor of 10,000. The interaction of such intense lasers with matter produces highly relativistic electrons and a host of nonlinear plasma effects. For example, a laser beam can carve out a channel in a plasma, keeping itself focused over long distances. Other nonlinear effects are being harnessed to produce novel accelerators and radiation sources. These high-intensity lasers are also used to produce shocks and hydrodynamic instabilities similar to those expected in supernovae and other astrophysical situations.

High-energy-density physics often requires large facilities. The National Ignition Facility (NIF) at the Lawrence Livermore National Laboratory will study the compression of tiny pellets containing fusion fuel (deuterium and tritium) with a high-energy laser. These experiments are important for fusion energy and astrophysics. During compression, which lasts a few billionths of a second, the properties of highly compressed hydrogen are measured. The information gained is directly relevant to the behavior of the hydrogen in the center of stars. The NIF will also study the interaction of intense x rays and matter and the subsequent shock-wave generation. These topics are also being studied at the Sandia National Laboratories using x rays produced by dense plasma discharges. A commercial source of fusion energy by the compression of pellets will require developing an efficient driver to replace the NIF laser.

PHYSICS IN BIOLOGY

The intersection between physics and biology has blossomed in recent years. In many areas of biology, function depends so directly on physical processes that these systems may be fruitfully approached from a physical viewpoint. For example, the brain sends electrical signals (i.e., nerve impulses), and these signaling systems are properly viewed as a standard physical system (that happens to be alive). In this section, the committee describes several important areas at the interface of physics and biology and discusses some future directions.

Structural Biology

Modern structural biology, that branch of biophysics responsible for determining the structure of biological molecules down to the position of the individual constituent atoms, had its origins in the work on x-ray crystallography by von Laue (Nobel Prize in 1914) and W.H. Bragg and W.L. Bragg (Nobel Prize in 1915). This was followed by the discovery of nuclear magnetic resonance by Rabi and by Purcell and Bloch, resulting in Nobel Prizes in 1944 and 1952. With x-ray crystallography and modern computers, the structure of any molecule that forms crystals can be determined down to the position of all of the individual atoms. Through technological advances depending on microelectronics and computers, it is now possible to determine the detailed structure of many protein molecules in their natural state. The structures of hundreds of proteins are now known, and biophysicists are learning how these proteins do their jobs. For many enzymes, the structural rearrangements and specific chemical interactions that underlie the enzymatic activity are now known, and how molecular motors move is becoming understood.

The main challenges that remain are determining how information is transmitted from one region of a molecule to another—from a binding site on one surface of a protein to an enzymatic active site on the opposite surface, for example—and how the specific sequence of amino acids determines the final structure of the protein. Proteins fold consistently and rapidly, but how they search out their proper place in a vast space of possible structures is not understood. Understanding how proteins assume their correct shape remains a central problem in biophysics (see sidebar "Protein Folding").

Biomechanics

Cells depend on molecular motors for a variety of functions, such as cell division, material transport, and motion of all or part of the organism (like the beating of the heart or the movement of a limb). At mid-century it was learned that muscles use specific proteins to generate force, but the

PROTEIN FOLDING

Proteins are giant molecules that play key roles as diverse as photosynthetic centers in green plants, light receptors in the eye, oxygen transporters in blood, motor molecules in muscle, ion valves in nerve conduction, and chemical catalysts (enzymes).

Their architecture is derived from information in the genes, but this information gives us the rather disappointing one-dimensional chain molecule seen at left in the figure. Only when the chain folds into its compact "native state" does its inherent structural motif appear, which enables guessing the protein's function or targeting it with a drug.

How can this folding process be understood? Experimentation gives only hints, so computation must play a key role. In the computer it is possible to calculate the trajectories of the constituent atoms of the protein molecule and thus simulate the folding process. This computation of the forces acting on an atom due to other protein atoms and to the atoms in the surrounding water must be repeated many times on the fast time scale of the twinkling atomic motions, which is 10^{-15} seconds (1 fs). Since even the fastest-folding proteins take close to 100 μs (10^{-4} s) to fold, all the forces must be computed as many as 10^{11} times. The computational power necessary to do this is enormous. Even for a small protein, a single calculation of all the forces takes 10^{10} computer operations, so to do this 10^{11} times in, say, 10 days requires on the order of 10^{15} computer operations per second (1 petaflop). This is 50 times the entire computational power of the world's population of supercomputers.

Even though the power of currently delivered supercomputers is little more than 1 teraflop (10^{12} operations per second), computers with petaflop power for tasks such as protein folding are anticipated in the near future by exploiting the latest silicon technology together with new ideas in computer design. For example, IBM has a project named Blue Gene, which is dedicated to biological objectives such as protein folding and aims at building a petaflop computer within 5 years.

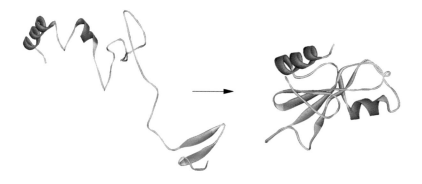

Unfolded State   **Native State**

multitude of motor types now known was only dimly suspected. Large families of motors have now been characterized, and the sequence of steps through which they exert force has been largely elucidated. In recent years researchers have measured the forces generated by a single motor cycle and the size of the elementary mechanical step. This research has required the invention of new kinds of equipment, such as optical tweezers, that permit precise manipulation of single bacterial cells as well as novel techniques for measuring movements that are too small to see with an ordinary microscope.

The challenge for the early part of the 21st century will be to develop a theory that relates force and motion generation with the detailed structure of the various types of motor proteins. The end result will be an understanding of how the heart pumps blood and how muscles contract.

Photobiology

Photosynthesis in plants and vision in animals are examples of photobiology, in which the energy associated with the absorption of light can be stored for later use (photosynthesis) or turned into a signal that can be sent to other places (photoreception). The parts of molecules responsible for absorbing light of particular colors, called chromophors, have long been known, as have the proteins responsible for the succeeding steps in the use of the absorbed energy. In recent years biophysicists have identified a sequence of well-defined stages these molecules pass through in using the light energy, some lasting only a billionth of a second. Biophysicists and chemists have learned much about the structure of these proteins, but not yet at the level of the positions of all atoms, except for one specialized bacterial protein, the bacterial photosynthetic reaction center (the work of Deisenhofer, Huber, and Michel was rewarded with the Nobel Prize in 1998).

In photobiology, the challenge is to determine the position of all atoms in the proteins that process energy from absorbed light and to learn how this energy is used by the protein. This will involve explaining, in structural terms, the identity and function role of all of the intermediate stages molecules pass through as they do their jobs.

Ion Channels

The surface membranes that define the boundaries of all cells also prevent sodium, potassium, and calcium ions, essential for the life processes of a cell, from entering or leaving the cell. To allow these ions through in a controlled manner, a cell's membrane is provided with specialized proteins known as ion channels. These channels are essential to the function of all cell types, but are particularly important for excitable tissues such as heart, muscle, and brain. All cells use metabolic energy (derived from the food we eat) to maintain sodium, potassium, and calcium ions at different concentrations inside and outside the cell. The special proteins that maintain these ionic concentration differences are called pumps. As a result of pump action, the potassium ion concentration is high inside cells and the sodium ion concentration is low: Potassium ions are inside the cell and sodium ions are outside. Various ion channels separately regulate the inflow and outflow of the different kinds of ions (letting sodium come in or permitting potassium to leave the cell). These channels working in concert produce electrical signals, such as the nerve impulses used for transmitting information rapidly from one place in the brain to another or from the brain to muscles.

Ion channels exhibit two essential properties, gating and permeation. "Gating" refers to the process by which the channels control the flow of ions across the cell membrane. "Permeation" refers to the physical mechanisms involved in the movement of ions through a submicroscopic trough in the middle of the channel protein: A channel opens its gate to let ions move in and out of the cell. Permeation is a passive process (that is, ions move according to their concentration and voltage gradients) but a rather complicated one in that channels exhibit great specificity for the ions permitted to permeate. Some channel types will allow the passage of sodium ions but exclude potassium and calcium ions, whereas others allow potassium or calcium ions to pass. The ability of an ion channel to determine the species of ion that can pass through is known as "selectivity."

At mid-century, it was realized that specific ion fluxes were responsible for the electrical activity of the nervous system. Soon, Hodgkin and Huxley provided a quantitative theory that described the ionic currents responsible for the nerve impulse (in 1963 they received the Nobel Prize for this work), but the nature of the hypothetical channels responsible remained mysterious, as did the physical basis for gating and selectivity.

In the past half-century, ion channels have progressed from being hypothetical entities to well-described families of proteins. The currents that flow through single copies of these proteins are routinely recorded (in 1991, Neher and Sakmann received the Nobel Prize for this achievement), and such studies of individual molecular properties, combined with modern molecular biological methods, have provided a wealth of information about which parts of the protein are responsible for which properties. Very recently, the first crystal structure of the potassium channel was solved, with its precise structure having been determined down to the positions of its constituent atoms.

The physics underlying gating, permeation, and selectivity is well understood, and established theories are available for parts of these processes. The main challenge for channel biophysicists now is to connect the physical structure of ion channels to their function. A specific physical theory is needed that explains how, given the structure of the protein, selectivity, permeation, and gating arise. Because the channels are rather complicated, providing a complete explanation of how channels work will be difficult and must await more detailed information about the structure of more types of channels. This is the goal of channel biophysics in the early part of the 21st century. The end result will be an understanding, in atomic detail, of how neurons produce the nerve impulses that carry information in our own very complex computer, the brain.

Theoretical Biology and Bioinformatics

Theoretical approaches developed in physics—for example, those describing abrupt changes in states of matter (phase transitions)—are helping to illuminate the workings of biological systems. Three areas of biology in which this approach of physics is being applied fruitfully to improve understanding are computation by the brain, complex systems in which patterns arise from the cooperation of many small component parts, and bioinformatics. Although neurobiologists have made dramatic progress in understanding the organization of the brain and its function, further advances will likely require a more central role for theory. To date, most progress has been made by guessing, based on experiments, how parts of the brain work. But as the more complicated and abstract functions of the brain are studied, the chances of proceeding successfully without more theory diminish rapidly. Because physics offers examples of advanced and successful uses of theory, the methods used in physics may be most appropriate for understanding brain structure and function.

The life of all cells depends on interactions among the genes and enzymes that form vastly complicated genetic and biochemical networks. As biological knowledge increases, particularly as the entire set of genes that constitutes the blueprint for organisms—including humans—becomes known, the most important problems will involve treating these complex networks. Modern biology has provided means to measure experimentally the action of many—in some cases all—of an organism's genes in response to environmental changes, and new theories will be required for interpreting such experiments.

One of biology's most fundamental problems is explaining how organisms develop: How do the 10^5 genes that specify a human generate the correct organs with the right shape and properties? This question of pattern specification by human genes has strong similarities with many questions asked by physicists, and physics offers good models for approaching it and similar questions in biological systems.

The information explosion in biology produces vastly complex problems of data management and the extraction of useful information from giant databases. Methods from statistical physics have been helpful in approaching these problems and should become even more important as the volume of data increases.

EARTH AND ITS SURROUNDINGS

As the impact of humans on the environment increases, there is a greater need to understand Earth, its oceans and atmosphere, and the space around it. The economic and societal consequences of natural disasters and environmental change have been greatly magnified by the complexity of modern life. For this reason, the quest for predictive capability in environmental science has become a huge multidisciplinary scientific effort, in which physics plays an important role. In this section, the committee discusses three important areas—earthquake dynamics, coastal upwellings, and magnetic substorms—where new ideas are leading to dramatic progress.

Earthquakes

The 1994 Northridge earthquake (magnitude 6.8) in southern California was the single most expensive natural disaster in the history of the United States. Losses of $25 billion are estimated for damaged structures and their contents alone. Earthquakes are becoming more expensive with each year, not because they are happening more often but because of the increased population density in our cities. An earthquake on the Wasatch fault in Utah would have had little effect on the agrarian society in 1901; in 2001, a bustling Salt Lake City sprawls along the same fault scarp that raised the mountains that overlook it.

Earthquakes happen when the stress on a fault becomes large enough to cause it to slip. Releasing the stress on one fault often increases the stress on another fault, causing it to slip and creating an avalanche effect. Recent studies of computer models of coupled fault systems under stress have yielded important insights. A large class of these models produces a behavior called self-organized criticality, in which the model's fault system sits close to a point of only marginal stability. The stress is then released in avalanche events. A remarkably large number of models yield the same statistical behavior despite differences in microscopic dynamics. These models correctly predict that the frequency of earthquakes in an energy band is proportional to the energy raised to a power. This, in turn, allows the accurate prediction of seismic hazard probabilities.

Studies of natural earthquakes provide the opportunity to observe active faults and to understand how they work. The Navy's NAVSTAR GPS has provided a relatively inexpensive and precise method for measuring the movement of the ground around active faults. An exciting new development in recent years has been the use of synthetic aperture radar (SAR) to produce images (SAR interferograms) of the displacements associated with earthquake rupture.

The largest earthquakes on Earth occur within the boundary zones that separate the global system of moving tectonic plates. On most continents, plate boundaries are broad zones where Earth is deformed. Several active faults of varying orientations typically absorb the motion of any given region, raising mountains and opening basins as they move. Modeling suggests that the lower continental crust flows, slowly deforming its shape, while sandwiched between a brittle, earthquake-prone upper crust and a stiff, yet still moving, mantle. This ductile flow of the lower crust is thought to diffuse stress from the mantle portion of the tectonic plate to the active faults of the upper crust. The physical factors that govern this stress diffusion are still poorly understood (see sidebar "Pinatubo and the Challenge of Eruption Prediction").

Coastal Upwellings

Environmental phenomena involve the dynamic interaction between biological, chemical, and physical systems. The coastal upwellings off the California coast provide a powerful example of this interaction. Upwellings are cold jets of nutrient-rich water that rise to the surface at western-facing coasts. These sites yield a disproportionately large amount of plankton and deposit large amounts of organic carbon into the sediments. Instant satellite snapshots of quantities such as the ocean temperature and the chlorophyll content (indicating biological species) provide important global data for

PINATUBO AND THE CHALLENGE OF ERUPTION PREDICTION

Volcanoes consist of a complex fluid (silicate melt, crystals, and expansive volatile phases), a fractured "pressure vessel" of crustal rock, and a conduit leading to a vent at Earth's surface. Historically, eruption forecasting relied heavily on empirical recognition of familiar precursors. Sometimes, precursory changes escalate exponentially as the pressure vessel fails and magma rises. But more often than not, progress toward an eruption is halting, over months or years, and so variable that forecast windows are wide and confidence is low.

In unpopulated areas, volcanologists can refine forecasts as often as they like without the need to be correct all the time. But the

fertile land around volcanoes draws large populations, and most episodes of volcanic unrest challenge volcanologists to be specific and right the first time. Evacuations are costly and unpopular, and false alarms breed cynicism that may preclude a second evacuation. In 1991, the need to be right at the Pinatubo volcano in the Philippines was intense, yet the forecast was a product of hastily gathered, sparse data, various working hypotheses, and crossed fingers. Fortunately for evacuees and volcanologists alike, the volcano behaved as forecast.

As populations and expectations rise, uncertainties must be trimmed. Today's eruption forecasts use pattern recognition and logical, sometimes semiquantitative analysis of the processes that underlie the unrest, growing more precise and accurate. Tomorrow's sensors may reveal whole new parameters, time scales, and spatial scales of volcanic behavior, and tomorrow's tools for pattern recognition will mine a database of episodes of unrest that is now under development. And, increasingly, forecasts will use numerical modeling to narrow the explanations and projections of unrest. Volcanoes are one of nature's many complex systems, defying us to integrate physics, chemistry, and geology into ever-more-precise forecasts. Millions of lives and billions of dollars are at stake.

models that have to describe the physical (turbulent) mixing of the up-welling water with the coastal surface water, the evolution of the nutrients and other chemicals, and the population dynamics of the various biological species. Since the biology is too detailed to model entirely, models must abstract the important features. The rates of many of the processes are essentially determined by the mixing and spreading of the jet, which involves detailed fluid mechanics. Research groups have begun to understand the yearly variations and the sensitivities of these coastal ecosystems.

Magnetic Substorms

The solar wind is a stream of hot plasma originating in the fiery corona of the Sun and blowing out past the planets. This wind stretches Earth's magnetic field into a long tail pointing away from the Sun. Parts of the magnetic field in the tail change direction sporadically, connect to other parts, and spring back toward Earth. This energizes the electrons and produces spectacular auroras (the Northern Lights is one example). These magnetic substorms, as they are called, can severely damage communication satellites. Measurements of the fields and particles at many places in the magnetosphere have begun to reveal the intricate dynamics of a substorm. The process at the heart of a substorm, sporadic change in Earth's magnetic field, is a long-standing problem in plasma physics. Recent advances in understanding this phenomenon have come from laboratory experiments, in situ observation, computer simulation, and theory. It appears that turbulent motion of the electrons facilitates the rapid changes in the magnetic field—but just how this occurs is not yet fully understood.

3

Structure and Evolution of the Universe

Physics at the earliest moments of our universe, when it was unimaginably hot and dense, is intimately related to physics at the highest energies. The universe then, a fraction of a second after the Big Bang, was a hot quantum soup of the fundamental particles being revealed in today's particle accelerators. All the forces of nature were at play. Even gravity, a tiny force for the physics of individual atoms and nuclei today, was strong. The physics of that era set the course for the future evolution of the universe, leading to what we see today: a vast, expanding cosmos of stars, galaxies, and unknown dark matter extending 13 billion light-years in all directions.

The observations revealing this picture of the early universe can be made only with sophisticated instruments using technology developed largely out of basic research on the physics of matter in the laboratory. For example, detectors based on ultrapure semiconductor chips, carried in satellites guided by radiotelemetry and atomic clocks, detect the light emitted from the hot Big Bang that has been propagating toward us for the last 13 billion years, much cooled by the expansion of the universe.

The link between basic physics and the world beyond Earth revealed by astronomy is also one of understanding:

• By applying the principles of physics learned in laboratories on Earth we can explain our observations of distant parts of the universe.
• Astrophysical observations are becoming increasingly important probes of frontier questions in fundamental physics. As the questions of physics evolve to very large and very small length scales, more and more phenomena that are important at these scales are to be found in space.

The increasing dependence of astrophysical observation on physics-based technology and the deepening ties between the fundamental laws of physics and the phenomena observed in the distant cosmos are the dominant themes in today's exploration of the universe.

NEW TOOLS: NEW WINDOWS ON THE UNIVERSE

Advances in physics are crucial for the development of new observational tools that further our understanding of the cosmos. New tools that measure electromagnetic wavelengths from radio waves to gamma rays have widened our window on the universe (see sidebar "Next Steps in the Exploration of the Universe"). And, using neutrinos and gravitational waves, new windows are being opened (see sidebar "Three New Windows").

NEXT STEPS IN THE EXPLORATION OF THE UNIVERSE

Our ability to study the universe is growing dramatically. We now view the universe with eyes that are sensitive to wavelengths from radio waves 10 cm long to gamma rays of wavelength 10^{-16} cm. Advances in materials and device physics have spawned a new generation of low-noise, high-sensitivity detectors. These new eyes have allowed us to see the universe as early as 300,000 years after its birth, to detect the presence of black holes and neutron stars, and to watch the birth of stars and galaxies. These photos show two of the next steps in this exploration: The Sloan Digital Sky Survey's 2.5-m telescope at Apache Point, New Mexico (left), will chart a million galaxies; the Chandra X-ray Observatory (right) will extend our view of the x-ray universe out past 5 billion light-years.

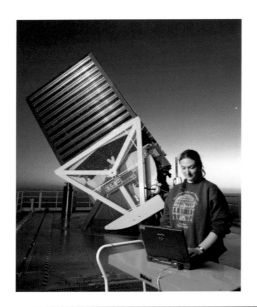

THREE NEW WINDOWS

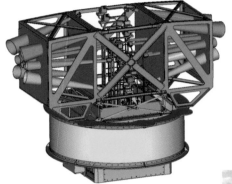

By 2003 the Microwave Anisotropy Probe satellite (MAP) (left) will produce unprecedented high-resolution images of the cosmic microwave background, the cooling fireball of the Big Bang, seen at a time when the universe was only one thousandth its present size. Together with maps of the elusive dark matter in our vicinity, these data will lead to a new era in precision cosmology.

LIGO (right) is a set of giant laser interferometers sensitive to ripples in the fabric of space-time. It may detect gravitational radiation, waves in the space-time warp produced by coalescing nearby pairs of neutron stars or black holes.

The BOREXINO neutrino detector (left) will for the first time allow us to see the elusive low-energy neutrinos from the decay of ^{7}Be in the Sun's core. This measurement will provide critical new data on the mass of the neutrino and its other properties.

Electromagnetic Waves

At radio wavelengths, giant interferometers employing radio telescopes spanning Earth chart the fall of matter into supermassive black holes. Millimeter-wavelength radio telescopes give us a view of astrophysical molecules and the cosmos at very early times. New satellite experiments map the microwave radiation that is the residual light from the Big Bang, giving us a detailed picture of the universe only 300,000 years after its beginning. Newly constructed large ground-based telescopes covering the infrared spectrum will be used to study galaxies at the time of their formation 11 billion to 13 billion years ago.

Optical telescopes are increasingly effective at capturing the feeble light from objects at the edge of the observable universe. The Hubble Space Telescope has vastly increased this reach, and the Next-Generation Space Telescope would probe the cosmos to even greater distances, seeing optical and infrared light produced at even earlier epochs. Wide-angle surveys probing the cosmos will generate terabytes of data and provide unique opportunities for understanding the cosmic forces generated by the dark matter that fills the universe. These surveys will also help us to understand how the arrangement of the galaxies developed.

At the shortest wavelengths, x-ray and gamma-ray studies of the universe are in their infancy but are already providing spectacular surprises. The Rossi X-ray Timing Explorer has tested general relativity in strong gravitational fields. The orbiting Chandra X-ray Observatory is pushing the exploration of x-ray luminous clusters of galaxies out to extreme distances and probing nearby supernova remnants for the elusive cause of their explosions. Satellite gamma-ray telescopes have revealed a class of ultraluminous sources of gamma radiation, some of which mysteriously emit a burst of radiation brighter than any other object in the universe. Several of these gamma-ray bursts have been associated with unusual supernovae. Satellite experiments are planned that will extend this exciting frontier, revealing more about these sources.

Neutrinos and Gravitational Waves

Neutrinos and gravitational waves offer new windows on the universe very different from those available using electromagnetic waves. Neutrinos are a natural probe of extremely hot and dense environments, where they are copiously produced and from which their weak interaction allows them to escape. Supernova 1987A was the explosive death of a massive star; its burst of neutrinos was detected in the United States by the IMB (Irvine-Michigan-Brookhaven) experiment in northern Ohio and at the Kamiokande facility in the Japanese alps. The process of nuclear burning at the center of the Sun, which makes the Sun a star, also produces neutrinos. Sensitive detectors including Super-Kamiokande in Japan, SAGE in Russia, and GALLEX in Italy have confirmed that neutrinos come from the Sun at a level below expectations. This discrepancy is now thought to be due to new physics—the nonzero mass of neutrinos allowing for the mixing of neutrino species—and is a frontier area of particle physics. New detectors (SNO in Canada and BOREXINO in Italy) will soon provide critical tests of this idea.

Gravitational waves are ripples in space and time propagating with the speed of light. Predicted by Einstein's general theory of relativity, gravitational waves produced by mass in motion can be detected by observing the motion of test masses as the ripples pass by. But since gravity is the weakest of the forces that act on matter, gravitational waves require large receivers for their detection. While there is direct evidence for the existence of gravitational waves in the motion of compact stars, such waves have never been detected on Earth. The worldwide network of gravitational wave detectors now under construction will attempt to detect these minute ripples. The U.S. effort—the Laser Interferometer Gravitational-wave Observatory (LIGO)—consists of two 4-km-long interferometers in Hanford, Washington, and one in Livingston, Louisiana.

These detectors have a dual role. They are experiments that test fundamental physics: the existence and character of gravitational waves. But they also offer a new way to see astronomical phenomena. Gravity's weak coupling to matter makes gravitational waves a unique window on the universe. Once produced, very little of a wave is absorbed. Gravitational wave detectors may enable us to see deeper into the environment around massive black holes and to moments in the universe earlier than those accessible by electromagnetic radiation.

Neutrino and gravitational wave detectors, looking outward, are not the only new windows on the universe. There is compelling evidence that up to 90 percent of the matter in the universe is made up not of the familiar protons, neutrons, and electrons that are the building blocks of the stars and planets but of unknown particles. This is called "dark matter," as it is matter not found in luminous stars (see sidebar "Dark Matter Then and Now"). Can we detect such dark matter particles in the laboratory? The vast majority of the dark matter must be moving slowly to avoid disturbing the formation of galaxies. Weakly interacting, massive particles (WIMPs) are a possible candidate. If these make up the dark matter, Earth is drifting through a sea of

DARK MATTER THEN AND NOW

COBE map. Small temperature fluctuations of the microwave background, the cooling cosmic fireball, are seen in this map (enclosed by text) made by the Cosmic Background Explorer satellite. These temperature fluctuations are related to developing dark matter densities at a time when the universe was only one thousandth its current age. New satellites will give even higher-resolution snapshots of the early universe. Different candidate dark matter particles will cluster in different ways in the intervening billions of years.

Cosmic mirage. A huge concentration of dark matter—seen 10 billion years after the Big Bang—is revealed by the space-time warp it creates around its host cluster of galaxies (see image at bottom). This mass warps the images of background galaxies. Millions of these images may be analyzed to reconstruct a map of the mass distribution of the dark matter over large areas on the sky. This distribution, in turn, may be used to test theories of dark matter and structure formation, as well as to understand the underlying physics.

them as it travels around the Sun and as the Sun orbits the galactic center. Earth itself is therefore a moving platform for detectors of such dark matter. Several very sensitive experiments for direct laboratory detection are now under way. Other dark matter candidates, called axions, are also being sought in Earth-bound experiments.

All these advances in observational capability—the product of basic and applied research in the physics of materials, optics, and devices—are enabling us to explore the universe to its furthest reaches, to its earliest moments, and through its most explosive events.

NEW LINKS

In astrophysics, the basic laws of physics are used to understand the large variety of objects that can be seen in the universe (planets, stars, galaxies, black holes, gravitational waves and lenses, dark matter, pulsars, quasars, x-ray sources, and gamma-ray bursts, to name just a few). As the frontiers of physics move to larger and smaller length scales, astrophysics is becoming more strongly linked to the physics studied in laboratories on Earth. Three examples illustrate this trend: cosmology, nuclear astrophysics, and black holes.

Cosmology

We live in an evolving universe filled with tens of billions of galaxies within our sphere of observation. Cosmic structures, from galaxies to clusters of galaxies to superclusters to the universe itself, are mingled together with invisible dark matter whose presence is known only through its gravitational effects. The light received from the most distant galaxies takes us back to within a few billion years of the beginning. The microwave echo of the Big Bang is a snapshot of the universe long before galaxies formed. A multitude of observations over the past decade have permitted cross-checks of our basic model of the past universe as a dense, hot environment in which structure forms via gravitational instability driven by dark matter.

What is the nature of the dark matter in the universe? What controls the development of structure in our universe? What do we know about the geometry and topology of our universe? How did it originate? These are fundamental questions that can be addressed by observation and careful deduction from our laboratory-tested knowledge of physics.

Our understanding of the universe has increased dramatically in the last decade. We can now map the small temperature variations (30 parts per million) in the remnant microwave radiation from the epoch when electrons combined with protons, only 300,000 years after the Big Bang. These variations, in turn, are related to the underlying gravitational effects of dark mass-energy fluctuations left over from an even earlier time. Thus, it is possible to see a filtered version of the primordial universe—an important frontier for physics at energies higher than those achievable in Earth-bound accelerators.

Important clues to this new physics are coming from observations of dark mass-energy, including its spatial distribution and the clumping of dark matter over cosmic time. Until recently we have had to rely on proxies for observations of dark matter—for example, the notion that luminosity is somehow related to mass and that luminosity might trace the dark matter. However, virtually all of the dark matter is nonluminous: Stars (and related material) contribute a tiny fraction (about 0.4 percent) of the mass required to halt the expansion of the universe, while the amount of matter known to be present from several estimates of its gravitational effects contributes about 100 times as much as stars. The dark matter cannot be made out of the familiar nuclear building blocks (protons, neutrons): Our understanding of the synthesis of light elements in the Big Bang, together with recent measurements of the primeval abundance of deuterium, reveals a universe in which ordinary matter (even nonluminous dead stars and dust) composes only 10 to 20 percent of the total mass. What is this dark matter and how is it related to the physics of the hot early universe? Some understanding of this will come from mapping the dark matter itself.

We can now map the dark matter over vast regions of space using the gravitational bending of light, seen as the distortion of images of distant galaxies. These telltale image distortions reveal the "invisible" foreground dark matter clumps, yielding two-dimensional maps of the dark matter distribution on the sky. Maps of the dark matter at various cosmic epochs will tell us the story of structure formation and constrain the physical nature of dark matter.

Is there cosmologically significant energy other than the dark matter? Using a certain type of supernova (type Ia) as a calibrated light source ("standard candle"), two groups—the Supernova Cosmology Project and the High-Z Supernova Team—both conclude that the expansion of the universe is accelerating rather than decelerating. If correct, this implies that much of the energy in the universe is in an unknown component, with negative pressure. The simplest explanation is a vacuum energy density (called a cosmological constant) with about 60 to 70 percent of the mass-energy required to make the universe flat. The rest would be made up by the dark matter itself.

Thanks to new tools, we are now entering the age of precision cosmology. When taken together, observations of the dark matter, dark energy, and fluctuations in the remnant radiation from the Big Bang will in the next few years give us a percent-level precision on several critical cosmological parameters, testing the foundations of our understanding of the universe. Because of the profound relationship between physics at the smallest distance scales and the details of the early universe and dark mass-energy, this will open a new window for physics.

Cosmology touches fundamental physics in many different ways. In elementary-particle physics, a unified theory of the basic constituents of matter is a major objective. If new particles make up the dark matter, they must fit naturally into such a scheme; conversely, unified theories predict the nature of the dark matter. In this way, cosmology and fundamental-particle physics are linked. If the missing energy is an energy of the vacuum (a cosmological constant), then the link is even stronger. The value of the cosmological constant is one of the major unsolved problems in the theory of the basic interactions. Simple estimates would lead to values as much as 10^{120} times greater than allowed by cosmological data. A definitive measurement of this central parameter from cosmology is critically important for the future development of particle physics.

Many features of the present universe can be explained by assuming that it underwent a very rapid expansion very early in its history. Inflation is the generic name given to this idea. Many different inflationary mechanisms have been proposed, all connected with the nature of the basic constituents of matter and how they behaved at that early epoch. Accelerator physics at the highest energies, which measures the properties of these constituents, explores the link between microphysics and cosmology. A nonrelativistic (cold) dark matter and inflation scenario is consistent with a large body of indirect observations: measurements of the anisotropy of the cosmic microwave background, redshift surveys of the distribution of luminous matter today, and supernova probes of acceleration. However, observations have only begun to discriminate between different inflationary mechanisms and different versions of cold dark matter.

The strong link between cosmology and elementary particle physics arises in part because the highest energies characterizing the particle physics frontier, unattainable in any conceivable accelerator on Earth, are actually reached in the Big Bang. The Big Bang thus provides one of the few "laboratories" for this kind of physics, although not one subject to our manipulation and control. Closer to home are the puzzling ultrahigh-energy cosmic rays that have recently been detected, a billion times more energetic than the particles in man-made accelerators. According to current understanding, such energetic cosmic rays should not exist: They should have lost their energy via interactions with the cosmic background radiation from the Big Bang. Do these cosmic rays portend new physics? We do not know, but if so it would not be the first time that cosmic rays have revealed new physics.

Gravitational physics is another area that is closely tied to cosmology. Gravity governs the structure and evolution of the universe on the largest possible scales. Although it is the weakest of the fundamental forces, it is not confined to small distances as are the nuclear forces, nor is it cancelled by opposite charges as is the electromagnetic force. It is universal, long range, and always attractive. Standard cosmological models incorporate the dynamics predicted by Einstein's general theory of relativity, as indeed does our whole idea of the Big Bang. In the future it should be possible to test Einstein's theory over these large distances by observing the changes in the expansion of the universe. Determining the cosmological constant is an important part of that test because the cosmological constant is one of the two parameters in Einstein's equation.

Gravitational physics is a two-frontier science, important on both the largest and smallest scales being researched by contemporary physics. The smallest scales, reached in the very early universe, are those of quantum gravity—discussed in Chapter 4—where our classical notions of space and time are expected to break down. Quantum gravity is important for cosmology not only because it is part of the fundamental dynamical theory describing the evolution of the universe but also because it frames another of the fundamental laws: a theory of the cosmological initial quantum state—a quantum theory of how the universe originated.

Nuclear Astrophysics

The physics of our universe is interwoven so that following one problem may lead to solving another. Recently, a combination of observations with the Beppo-SAX satellite and two ground-based optical telescopes produced proof that some sources of high-energy gamma-ray bursts are at high redshift and so must be the most luminous objects in the universe. Such events require energies that strain even the capabilities of black hole dynamics, and they could well be a source of measurable gravitational waves and, possibly, neutrino bursts. The possible identification of the unusual supernova SN1998bw with the equally special gamma-ray burst GRB980425 is a fascinating clue. Some supernovae produce energies 10 to

100 times larger than previously recognized and/or produce relativistic jets pointing at Earth. What is the nature of these gigantic explosions?

A first step in solving the puzzle is to simulate the explosions, using the consequences of known physics, and compare simulations with observations. The dramatic evolution in computer power is beginning to make such three-dimensional, time-dependent modeling feasible. An important issue is how such complex calculations may be tested for reliability in a way that is independent of the observations to be understood.

Some help may come from a surprising source. High-intensity lasers built for the study of inertial confinement fusion and for national defense projects have been used to make scale models of a supernova. This is done by blasting a small target with an enormous pulse of laser light. Although the laser targets have dimensions of 0.1 cm and explode in 20 ns, this is a good model for supernova mantles with dimensions larger than the Sun and durations of hours. Furthermore, petawatt lasers have reached such high intensities that they can produce relativistic plasmas, making possible the experimental study of relativistic jets of particles believed associated with several exotic astrophysical phenomena.

As in the Big Bang, explosive nucleosynthesis (the production of atomic nuclei in supernova explosions) is a diagnostic of the explosion itself. Utilizing this diagnostic requires understanding the nuclear microphysics underlying the supernova explosions, including the associated nucleosynthesis and its impact on the chemical evolution of our galaxy. Included in this will be the resolution of still unsolved problems in our understanding of the mechanism of supernova explosions and the origin of the heavy elements. Both, in turn, are connected to neutrino physics and the properties of exotic, neutron-rich nuclei, subjects that we hope to be able to study in the laboratory in the decade ahead. The modification of matter by thermonuclear alchemy using high-intensity lasers, and its subsequent ejection into conditions in which it may be analyzed, gives a means of probing these explosive environments.

This is another case of an intersection of two fields of physics (nuclear physics and astrophysics) that will receive an enormous push because of the booming technology of astrophysics: Hubble Space Telescope measurements of abundances in metal-poor stars, detection of gamma-ray lines from new sources (such as ^{44}Ti), x-ray pictures of supernova remnants, and nuclear chemistry probes of individual stellar grains. There is great potential for learning new nuclear astrophysics in the coming decade. The mechanisms for supernovae of type Ia (the standard candles mentioned earlier) and type II (those core-collapse supernovae that produce neutron stars and black holes) are two of the major computational challenges in nuclear astrophysics (see sidebar "Stars in the Laboratory"). These mechanisms may be related to the even more energetic explosions causing the gamma-ray bursts. These bursts, in turn, must also be accurately simulated; they are a

STARS IN THE LABORATORY

Most of the visible matter in the universe is in the form of yellow stars like the Sun. Their central temperatures and densities are becoming accessible to experiment by focusing light from many lasers into a tiny volume. Thermonuclear reactions may occur under these conditions, making these experiments interesting for research in astronomy, controlled fusion, and nuclear weapons.

The computer simulation of the interior of the explosion of supernova 1987a predicted complex instabilities between layers of the star (below).

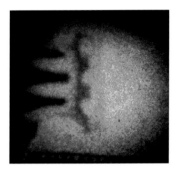

Experiments on the Omega laser at the University of Rochester provide small-scale versions of the explosion (100 billion times smaller in radius); above, right, are shown the effects of such instabilities in multiple layers. In this sense, the experiments give a scale

model of the supernova, on which both our ideas about the explosion and our computer simulations of it can be tested.

Constructing reliable and accurate simulations of such complex phenomenon is at the frontier of astrophysics and plasma physics. Below, a researcher shows his son how he simulates supernovae and laser experiments.

promising source of gravitational waves for the LIGO detectors. Finally, those same type Ia supernovae, which were used to infer the acceleration of the universe, must be understood better before they can provide an accurate distance scale. This will require an improved knowledge of the atomic physics of ionized cobalt and other iron-group elements.

Black Holes

Perhaps nowhere is the connection between fundamental physics and astrophysics clearer than in the physics of black holes. The key idea of Einstein's battle-tested theory of gravity—general relativity—is that mass curves four-dimensional space and time. General relativity predicts that when mass is compressed into a small enough volume, this curvature, and the consequent gravitational pull at the surface, will be too large for anything to escape, even light. The surface of a black hole, called its horizon, is like a one-way membrane. Matter, information, and observers can fall into it but nothing can emerge from it.

Black holes are surprisingly simple objects. Typically produced in nature by complex gravitational collapse, general relativity predicts that in the end they are characterized by just a few numbers. As S. Chandrasekhar put it, "The black holes of nature are the most perfect macroscopic objects there are in the universe: the only elements in their construction are our concepts of space and time. And since the general theory of relativity provides only a single unique family of solutions for their description, they are the simplest objects as well."[1] Black holes occur throughout the universe because the attractive force of gravity leads to continual compression of mass. Black holes with masses comparable to that of our Sun have been discovered all over our galaxy as unseen members of certain x-ray-emitting double star systems. These x rays are generated when hot gases are stripped from the normal companion star and fall into the black hole.

Supermassive black holes have been detected at the center of many galaxies, including our own, by observing their effects on the motion of the surrounding matter. These black holes are up to a billion times more massive than the Sun and are likely to be in the center of every large galaxy. Matter falling into such a huge black hole, which becomes hot and radiates as its pressure rises, is thought to be the source of the activity seen in active galactic nuclei (AGN) and the hugely luminous quasars. Astrophysicists are

---

[1]S. Chandrasekhar. 1992. *A Mathematical Theory of Black Holes.* Oxford University Press, New York, p. 1.

trying to understand the behavior of matter, radiation, and magnetic fields inside these accretion disks of infalling matter.

Present observations reveal black holes only indirectly—by their effects on surrounding luminous matter. Sufficiently advanced gravitational wave detectors should be able to receive directly the gravitational radiation from perturbed black holes and to test further the predictions of Einstein's theory. Making these predictions involves computation on very large scales, which in turn requires an intensive effort to formulate the computation effectively.

The massive black holes of astrophysics may not be the only black holes that exist in the universe. Microscopic black holes may have been produced in the early universe, small enough that quantum gravity could be important for their understanding. Although prohibited classically, matter inside black holes can tunnel out quantum mechanically. This effect, predicted by Stephen Hawking in the early 1970s, has the consequence that black holes emit radiation like blackbodies. Indeed, the smaller the mass, the higher the temperature, so that evaporating black holes eventually explode. In these explosions, which have not yet been detected experimentally, space-time curvatures are reached that are greater than any since the Big Bang. At these curvatures, space and time behave quantum mechanically, and physics at the highest energy scales, such as the physics associated with the unification of the fundamental interactions, plays an important role.

QUESTIONS AND OPPORTUNITIES

Our knowledge of the universe is accelerating, driven by new observational results. Cosmology is in the midst of a golden age. Within 10 years we may have a cosmological theory that explains almost all the fundamental features of the universe: its smoothness and flatness, the heat of the cosmic microwave background, the asymmetry between ordinary matter and so-called nonbaryonic matter, and the origin of structure. At the present time, however, there is much we do not understand.

Is the universe spatially flat, as predicted by inflation? Does the absence of antimatter and the tiny ratio of matter to radiation (around 10^{-10}) involve forces operating in the early universe that violate baryon-number conservation and matter-antimatter symmetry? Is inflation the correct theory of the earliest moments of the universe, and if not, what is? Is the expansion of the universe today accelerating rather than slowing down because of the presence of vacuum energy or something even more mysterious? How is this vacuum energy related to quantum gravity?

What are gamma-ray bursts? How do black holes form? Can we directly detect neutrinos from the relativistic jets in gamma-ray bursts? Can we detect gravitational waves from core collapse in a supernova or in a merger involving a black hole? What determines whether a core collapse becomes a black hole or a neutron star? The list goes on.

Exciting opportunities abound in the coming decade for exploring these fundamental questions, thanks to advances in applied physics and technology. A paradigm based on profound connections between cosmology and elementary particle physics—inflation along with cold dark matter—holds the promise of extending our understanding to an even more fundamental level and much earlier times and of shedding light on the unification of the forces and particles of nature. As we enter the 21st century, a flood of observations is testing this paradigm.

4

Fundamental Laws
and Symmetries

The observation of nature, be it with human eyes and ears or with particle accelerators and electron microscopes, often reveals distinctive patterns: Hurricanes recur every fall, sunspot activity peaks every 11 years, salt forms a cubic crystal, the chemical properties of atoms lie neatly in a periodic table. Physics attempts to bind these different phenomena together by searching for common underlying laws. The laws of magnetism, which explain a child's bar magnet, apply as well to the nuclear cores of atoms. The laws of gravitation, which govern the gentle drop of falling leaves, also determine the behavior of the black hole in the center of our galaxy.

In the early part of the 20th century, research in the subatomic realm revealed a simple, economical structure—all matter was composed of three elementary particles: the electron, the proton, and the neutron. Over the course of the second half of that century, probes at the even shorter distances of the nuclear and subnuclear realm led to several important changes in this picture. Early on, a new particle was added to the list: the neutrino, observed as an end-product in nuclear decay. This nearly weightless particle plays an essential role in nuclear decay, the lives of stars, and even, possibly, the ultimate fate of our universe. Nevertheless the neutrino remains mysterious, with ongoing experiments in Japan, North America, and Europe only beginning to elucidate its nature. In the 1960s the proton and neutron were revealed to have substructure: They contain two species of yet smaller particles called quarks. The two species are known as the up quark and the down quark. Most of the matter we encounter is composed of these four particles alone. But, strangely, nature has chosen to replicate this list (up quark, down quark, electron, and its neutrino) twice more, with the two additional sets differing only by being much heavier. The last entry in this catalog, the top quark, with a weight more than 30,000 times that of the up quark, was discovered only recently in the Tevatron collider at Fermilab.

As research has probed the constituents of matter, it has also revealed the interactions among them. The gravitational force described by Newton in the 17th century dominates physics at the largest distances of our universe. The electromagnetic force, synthesized triumphantly in the 19th century in the form of Maxwell's equations, determines the behavior of atoms, molecules, and materials. Investigation of the subatomic world has so far revealed two further forces that operate only at the most minute distances: the strong force, responsible for the structure of nuclei, and the weak force, first revealed in nuclear decay. The constituents of matter together with these four forces constitute the underlying structure of contemporary physics, the so-called standard model.

HIDDEN SYMMETRIES AND THE STANDARD MODEL

One of the deep ideas embodied in the standard model is the notion that symmetries may be hidden—what we observe in nature may not directly reflect the underlying symmetries of the laws of physics. The spherically symmetric laws of electricity and magnetism, which determine most of the phenomena of our everyday experience, are equally responsible for the highly irregular snowflake and the round raindrop. In the first case the spherical symmetry of electricity and magnetism is hidden, while in the second it is manifest. Materials that do not exhibit the symmetry, like the ice in the snowflake, are in a different "phase" from materials like liquid water, in which the symmetry is evident. Frequently, changes in the environment with which the material interacts—the air temperature, the pressure, or other factors—determine whether the symmetry is hidden or revealed. Thus by heating the snowflake (hidden symmetry) we obtain the raindrop (manifest symmetry). This change of a substance from a hidden-symmetry phase to another where it is manifest, a phase transition, plays a role in nearly all branches of physics.

The idea of hidden symmetry is central to understanding the pattern of elementary particles and their interactions. One celebrated example, the relationship between radioactive decays and electromagnetic phenomena, was originally introduced as an analogy: Enrico Fermi suggested that the force responsible for certain radioactive decays, now known as the weak interaction, might be described in terms of weak charges and weak currents, just as electromagnetism involves ordinary charges and currents. The analogy was not precise—the analogue of the electromagnetic radiation was absent from Fermi's theory. Although Fermi's idea did partially explain the weak interaction, it was unable to encompass all of the phenomena observed in radioactive decay or those observed in the properties of "hadrons," bound states of quarks.

The Electroweak Force

In 1962, Sheldon Glashow proposed a connection between weak and electromagnetic interactions that was deeper than that of Fermi. In effect he unified electromagnetism and weak processes into a single electroweak interaction using an expanded version of the gauge invariance of electromagnetism. This symmetry had the immediate consequence that there should be a "weak" analogue of the electromagnetic field, with the weak force transmitted by new particles analogous to the photon of electromagnetism. Called the W and the Z, these particles were first observed at CERN, in Geneva, Switzerland, 20 years after Glashow first made his suggestion.

Although gauge invariance for the electroweak interaction had appealing features, it was not readily accepted when first proposed. The reason was that nature did not appear to possess this symmetry. For example, the symmetry requires the electroweak interaction to extend to large distances, a so-called long-range force. While the electromagnetic interaction is indeed long range, the weak forces are extremely short range, acting on distances much smaller than an atomic nucleus.

The solution to this puzzle grew out of efforts in the 1960s by both elementary-particle and condensed-matter physicists to investigate the consequences of hiding a symmetry. It was soon realized that unexpected phenomena can arise from hidden symmetries. For example, at very low temperatures, where the symmetries of electromagnetism become hidden, materials often display superconductivity: the absence of resistance to current flow, the expulsion of magnetic fields, and only short-range electromagnetic interactions. Study of these phenomena led Steven Weinberg to suggest that the symmetry of the electroweak interaction might be similarly hidden, with similar consequences. In particular, this hidden symmetry turned the long-range interaction into a short-range one. In this form the weak interactions are in a superconducting phase in which the symmetry is hidden. This final form of the electroweak theory has been beautifully confirmed through 30 years of accurate experimental tests.

The details of how the electroweak symmetry is hidden are mysterious. The unraveling of this physics remains one of the most important questions in particle physics. Weinberg's original suggestion for the physics responsible for hiding the electroweak symmetry requires a new object, the Higgs boson. As the missing player in the standard model, the Higgs has been the subject of intense searches at CERN's electron-positron collider (LEP) and at the Tevatron collider at Fermilab outside Chicago (see sidebars "Tools of the

Trade" and "Societal Benefits from Accelerator Science"). If the Higgs boson exists, the fact that it has not been observed implies that it weighs 110 times more than the proton! Early in 2001, experiments at a newly upgraded Tevatron will extend the Higgs search to even higher masses. CERN's Large

TOOLS OF THE TRADE

Work at the cutting edge of experimental physics requires instrumentation that pushes the frontiers of technology. Just as astronomy has seen amazing advances brought about by the latest generation of large-scale modern instruments, so too have the small distance frontiers of physics been advanced by sophisticated, large-scale facilities. In particle and nuclear physics, where the smallest structures in nature are probed, the increasing capability of particle accelerators, the modern "microscopes" designed for this purpose, along with their instrumentation, have driven the experimental frontier. From Ernest Lawrence's original 4-in. cyclotron to today's mammoth colliders—one of them, the Collider Detector at the Fermilab (CDF), is shown below at the left; below at the right is an illustration of a top quark decaying in the CDF—the energy of particle beams has increased by a factor of over a million. At the same time, a series of breakthroughs has decreased the cost per unit energy, keeping the high-energy frontier within the range of economic feasibility.

In the new century, the suite of accelerators and detectors available to the nuclear and high-energy physics community will provide answers to the most important questions in these fields. The growth of accelerator-based detectors from a modest array of Geiger counters and scintillators in the 1940s and 1950s to today's thousand-ton, house-size facilities has been remarkable. Huge detectors have powerful magnets enclosing arrays of hundreds of thousands of sensor elements, each feeding data at high speeds to sophisticated data acquisition and processing computers. New technologies adapted or developed for these detectors drive a rich interchange between the scientific community and segments of the manufacturing, electronics, and computer industries.

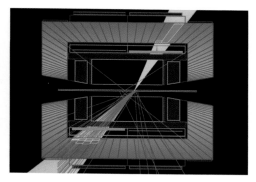

SOCIETAL BENEFITS FROM ACCELERATOR SCIENCE

The flowering of accelerator science and technology in response to the needs of nuclear and high-energy physics has spawned an array of technologies benefiting health care, industry, and basic research in many fields of science. A significant fraction of the radioisotopes used in medical treatment, diagnostics, and research are produced by accelerators. Beams from accelerators are successfully used in the treatment of cancer and other diseases. Modern medical imaging techniques such as CAT scans, PET scans, and MRI have their roots directly in technologies developed for particle detectors. The development of readout electronics and data acquisition technologies for these detectors has led to new industrial processes and products.

In industry, R&D utilizing accelerators is often undertaken to develop new products, for example high-density magnetic storage media. Beams are used to alter the composition of materials by means of techniques such as ion implantation; to improve materials performance (e.g., hardening surfaces to increase wear resistance); for direct process improvement (e.g., curing of epoxies and plastics); and directly in such production processes as x-ray micromachining.

Accelerators are used to produce intense, bright beams of x-rays and neutrons for forefront basic and applied research in the life sciences, chemistry, materials science, geology, and the environmental sciences. Large-scale accelerator-based facilities are operated for thousands of users from universities, industry, and the national laboratories. Programs in areas such as materials characterization, protein crystallography, surface characterization, and chemical dynamics are advancing the forefront of our knowledge.

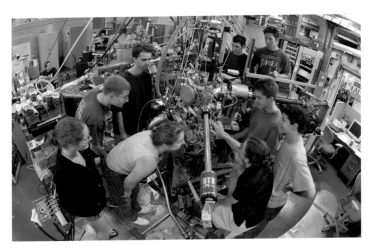

Students participate in experiments at the Advanced Light Source at the Lawrence Berkeley National Laboratory using an intense beam of x rays as a microscopic probe of materials and surfaces.

Hadron Collider (LHC), now under construction and expected to be completed in 2005, will produce collisions with an energy much larger than that available today, significantly advancing the energy frontier. There are strong arguments that the Higgs should lie within the LHC's range of sensitivity. The properties of the Higgs boson would be peculiar and unlike those of any known particle. For this reason, other suggestions have been made for the origin of electroweak symmetry breaking, all of which lead to predictions of phenomena whose signals will be observable at the Tevatron or the LHC.

Many other aspects of this standard model of electroweak forces, such as the strength of the various particle interactions, have been subjected to detailed tests at facilities like LEP and the SLAC electron-positron collider, know as SLC. These experiments are scientific marvels involving massive detectors and huge international scientific collaborations. The results are in spectacular agreement with the predictions of the theory. Important tests of the standard model have also been performed at low energies. A notable example is the contribution of the weak force to atomic structure. Because the electromagnetic interaction is so much stronger, the weak effects are very small, about 1 part in 100 billion. However, the weak interaction is not symmetric under mirror reflection; that is, it does not display parity. Experimenters recently exploited this parity violation in measuring the weak effects to an accuracy of better than 1 percent, an extraordinary achievement made possible by advances in laser technology. The results again confirm the standard model.

In superconducting materials, the superconductivity disappears as the temperature of the material is raised, revealing the electromagnetic symmetry. So, too, the electroweak symmetry should become manifest at sufficiently high temperatures. Although such temperatures have not been achieved in the laboratory, they were probably reached in the early universe, shortly after the Big Bang. Understanding how the electroweak symmetry is hidden will reveal details about the cosmology of our universe more than 13 billion years in the past.

The Strong Force

Hidden symmetry also plays a crucial role in the theory of the strong force. The underlying constituents of strongly interacting matter, the quarks, do not appear individually under normal conditions but only as composite bound states called hadrons. Protons and neutrons, described earlier as three-quark bound states, are examples of hadrons. The quarks within protons and neutrons are extremely light objects, yet protons and neutrons are hundreds of times heavier. This binding by the strong force, quantum chromodynamics (QCD), hides a symmetry associated with the nearly weightless quarks, allowing the proton and neutron to be heavy. One surprising consequence of this picture is the existence of a quark and an antiquark bound state much lighter than a proton or neutron. This particle, called the pion, was first observed in cosmic-ray interactions. The exchange of pions between nucleons is an important part of the mechanism by which the strong interaction binds nucleons into nuclei. The pion's small mass was a mystery resolved by the discovery of a strong-interaction hidden symmetry.

As in the electroweak interactions, the hidden quark symmetry may be restored at very high temperatures or very high energy densities. The newly commissioned Relativistic Heavy Ion Collider (RHIC) at Brookhaven National Laboratory on Long Island is designed to reach these extreme temperatures and densities, leading to the creation of a new state of quark matter (see sidebar "Recreating the Early Universe in the Laboratory"). A similar heavy ion experiment, ALICE, will be performed at CERN's LHC.

Obtaining predictions from the theory of strong interactions is often difficult because of the technical complexities of QCD, difficulties that do not arise for weaker interactions such as the electromagnetic force that binds electrons to the nucleus. In recent years, owing to the advent of powerful supercomputers, this situation is beginning to change. A technique known as lattice gauge theory can be used to calculate many properties of hadrons and may someday allow us to calculate directly from QCD the structure of the nucleus. It may also shed light on other aspects of "hadronic" physics, such as hadrons without quarks (called glueballs) and strange matter. These ideas are being tested experimentally at the Jefferson Laboratory in Newport News, Virginia.

CP Symmetry

Perhaps the most mysterious aspect of the standard model is the breakdown of a symmetry called CP. This symmetry involves a reversal of sign of all of a particle's charges and a simultaneous mirror reflection of space. Until the 1960s it was thought that CP was an exact symmetry of nature. But this notion was shattered with the discovery that the decay of a hadron called the K meson was slightly different from that of its antiparticle. Because the difference is very small, CP is nearly a perfect symmetry.

While the CP violation seen in K decays can be accommodated by the standard model, this model provides no insight into the origin of the CP

RECREATING THE EARLY UNIVERSE IN THE LABORATORY

One millionth of a second after the Big Bang, our universe was filled with a soup of quarks and gluons, the particles we believe to be the fundamental building blocks for the nuclear matter of our everyday world. Later, as the universe expanded and cooled, the quarks and gluons "froze" together to form the familiar protons and neutrons that make up atomic nuclei.

Theorists have argued that it should be possible to recreate the conditions of the Big Bang, and thus the primordial quark-gluon soup, if sufficiently extreme conditions of temperature and pressure are achieved. Experimentalists can now build powerful particle accelerators and colliders to produce these conditions, though only in a volume the size of a large atomic nucleus. The quark-gluon soup (or "plasma") is generated by using the energy of motion of colliding nuclei to heat them to a temperature where the individual neutrons and protons "melt," allowing their constituent quarks and gluons to roam freely throughout the nuclear volume.

At Brookhaven National Laboratory near New York City, a large accelerator called the Relativistic Heavy Ion Collider (RHIC) has been constructed to accomplish this goal. RHIC allows physicists to collide gold nuclei, head-on, with energies of 20 trillion eV per nucleus. (See below for an image of this collision as seen by the Star Detector at RHIC.) As of the writing of this report, the first data had been obtained and were being examined for evidence of the quark-gluon plasma.

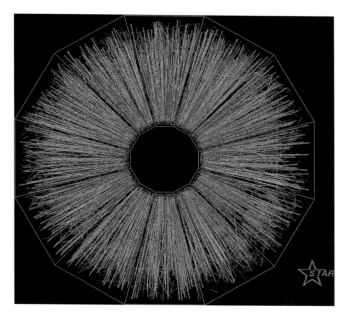

violation. A new generation of particle accelerators may change this situation. The so-called B-Factories at SLAC, Cornell University, and the KEK laboratory in Japan will produce large numbers of B mesons. These are similar to K mesons except that an s-quark, which belongs to the second family, is replaced by a third-family b-quark. By examining certain rare decays of the B meson, CP-violating quantities can be isolated. The important question is whether B meson decays can be explained with the same parameter introduced to explain K decays. A negative result will force us to extend the standard model to account for CP violation.

The standard model contains a second source of CP violation, known as strong CP violation, that is not probed in K or B decays but can be studied in low-energy atomic and nuclear experiments. This source of CP violation generates an electric dipole moment (EDM), a separation of charge along a particle's spin axis. If a particle with a nonzero EDM is placed in an electric field, its spin will precess around the direction of the field, just as a child's top precesses around the vertical. Wonderfully precise techniques have been developed to measure neutron and atomic EDMs. The former involve ultracold neutrons confined in "bottles." The latter employ atom beams, vapor cells, or atom traps. While no EDM has yet been detected, the precision of these experiments is extraordinary: Charge distortions of 1 part in 10 trillion trillion are ruled out! And experiments hundreds of times more sensitive are being planned. It is very difficult to understand why the standard model—which allows a nonzero EDM—does not generate one. One attractive solution of this strong CP puzzle involves a modification of the standard model's Higgs structure. This change has a remarkable consequence, a new and very light particle called the axion, which is a leading candidate for dark matter. Several experiments are under way to determine whether the universe is filled with a sea of axions.

Matter and Antimatter

The interest in CP violation is connected with a puzzling aspect of our universe—namely, that it contains much more matter than antimatter. If physics is manifestly CP symmetric at high temperatures, one would expect the very early universe to contain equal numbers of particles and antiparticles. If this symmetry had been maintained during the subsequent expansion and cooling, matter and antimatter would have almost exactly annihilated each other, leaving a vast void containing only a faint glow of radiation from the primeval fireball. Instead, we find ourselves surrounded by planets, stars, and other matter. It is known that one requirement for generating the observed matter-antimatter asymmetry during expansion is CP violation, and it is believed that the CP violation seen in K meson decays might not be sufficient. Thus the CP violation of the standard model may be just the first indication of the new physics that lurks beyond that model.

NEW PHYSICS FOR A NEW ERA

The standard model is a triumphant success. Yet for every question it answers, new and deeper ones arise. Why are there three families of quarks and leptons? What accounts for the patterns of masses within the families? Why, for example, is the electron neutrino at least 100,000 times lighter than its charged partner, the electron? Why are certain interactions mixing the families absent in the standard model? Why are the W and Z particles so heavy? Why are some symmetries exact in nature and others broken or hidden? Why are symmetries so essential to nature?

Ideas have emerged in the past few years that promise to answer these questions. Many of them involve the behavior of matter at increasingly smaller distance scales, distances requiring increasingly energetic probes. More powerful accelerators turning on early in this century will continue to push back this energy frontier. Some of the questions involve distance scales so small that they will probably never be probed directly in accelerator experiments. But subtle hints of this new physics may be encoded in phenomena like neutrino masses, free proton decay, and family mixing.

New Symmetries

There must be new forces responsible for the hidden symmetries of the standard model, forces of such short range that they have so far escaped discovery. But there are powerful arguments suggesting that the next generation of accelerators will be sufficiently energetic to produce the massive particles that carry these forces. While there are several appealing ideas about the nature of the physics that will be discovered, especially intriguing is an entirely new type of symmetry known as supersymmetry.

Supersymmetry predicts a new partner for each of the known particles. For example, the partners of the quark and photon are known as the "squark" and "photino," respectively. None of these particles has yet been seen, indicating that supersymmetry is hidden in such a way that these partners are so massive as to have escaped detection. But if supersymmetry is related to the hidden symmetry of the standard model, then these masses cannot be too large, and they can be produced in collider experiments. In particular, the search for supersymmetry will be central to the LHC program that begins in 2005. The observation of such superpartners and their associated supersymmetery would be a stunning experimental discovery.

Grand Unification

A puzzling feature of the standard model is the large disparity in strength between electroweak interactions and the strong interaction of QCD. Yet these interactions have very similar structures, both modeled on the gauge invariance of electromagnetism. Soon after the emergence of the standard model, it was realized and confirmed experimentally that the strengths of these interactions change with distance: As the distance gets smaller, the strong interaction grows weaker and the weak interaction grows stronger. They ultimately come together at distances roughly a million billion times smaller than those now being probed. At this tiny distance the strong and electroweak theories may combine into a single grand unified theory, incorporating symmetries beyond those of the strong and electroweak theories.

An especially dramatic prediction of grand unified theories is that the proton, a basic building block of matter, decays into lighter particles with a very small probability. In fact, a large class of grand unified theories predict that protons have a mean lifetime in excess of a trillion trillion times the age of our universe. This explains why we still have plenty of protons. Despite 20 years of careful searches with massive detectors located far underground to avoid spurious signals, no proton decay has been seen, ruling out the simplest grand unified theories. However, some of the most appealing versions of these theories incorporating supersymmetry predict that current detectors are very close to having the required sensitivity.

Neutrinos

A startling discovery made recently may be a different sign of grand unification. Nuclear and particle experimentalists using underground detectors similar to those involved in the search for proton decay detect neutrinos produced by the Sun and by cosmic-ray interactions in Earth's atmosphere (see sidebar "Massive Neutrinos and Neutrino Astrophysics"). The results indicate that a neutrino belonging to one family can spontaneously transform into a neutrino from another, a process known as neutrino oscillation. This can happen only if neutrinos have a nonzero mass, a possibility outside the standard model but easily accommodated in grand unified theories. In fact, the mass determined by the Super-Kamiokande atmospheric neutrino experiment in Japan may be a sign of new interactions at the grand unified scale. Housed deep within a mine in the Japanese alps, Super-Kamiokande contains 50,000 tons of ultrapure water, the inner portion of which is viewed by an array of 13,000 photomultiplier tubes.

MASSIVE NEUTRINOS AND NEUTRINO ASTROPHYSICS

Earth is bathed in an enormous flux of neutrinos produced by thermonuclear reactions occurring deep in the core of our Sun. The prospect of exploiting neutrinos as a solar probe led experimenters to build a series of neutrino detectors, all of which recorded fewer neutrino events than expected. Efforts to explain the results by modifying the standard solar model proved unsuccessful. Gradually physicists began to suspect that the explanation had to do with unexpected properties of neutrinos. If neutrinos have a mass, then the electron neutrinos produced in the solar core can "oscillate" into muon or tau neutrinos before reaching Earth, thereby escaping detection.

The Super-Kamiokande experiment, an enormous 50,000-ton water Cherenkov detector located in a mine deep within the Japanese alps, has recently measured both solar neutrinos and neutrinos produced in our atmosphere by cosmic-ray interactions. When the detector, shown at the left, is filled with water, these 13,000 photomultiplier tubes detect faint flashes of light caused by passing neutrinos. The atmospheric neutrino results have already produced dramatic evidence for oscillations, an apparent transmutation of muon neutrinos into tau neutrinos.

The Sudbury Neutrino Observatory (SNO), located nearly 2 km underground in a Canadian nickel mine, contains 1000 tons of heavy water. The deuterium in the heavy water allows the experimenters to measure both solar electron neutrinos and their oscillation products, the muon and tau neutrinos. The SNO experiment is now under way. A possible explanation for the Super-Kamiokande results attributes the needed neutrino masses to physics deeply hidden at 10^{15} GeV, an energy roughly a thousand billion times greater than that reached with the largest accelerators.

Neutrino oscillations in which the electron neutrinos produced in the Sun convert into muon- or tau-type neutrinos that cannot be seen in detectors on Earth may also account for the peculiar behavior of neutrinos coming from the Sun. Five underground experiments have detected the electron neutrinos, but the rate at which they arrive at Earth is nearly three times smaller than that predicted by models of solar neutrino production. While the evidence is indirect (neither muon nor tau neutrinos produced in the oscillation have been directly observed), the results are in excellent accord with the oscillation hypothesis.

A favorite theoretical interpretation is that the electron neutrino produced in the Sun oscillates to a muon neutrino with a mass of about 0.003 eV. Such oscillations would take place predominantly in the Sun's interior as the neutrinos pass from the high-density solar core to the surface. In this dense interior, oscillations can be magnified a thousandfold. This effect, predicted 15 years ago by Mikheyev, Smirnov, and Wolfenstein and known as the MSW effect, is analogous to phenomena explored in atomic physics 50 years earlier. A new solar neutrino detector that uses heavy water, the Sudbury Neutrino Observatory in Canada, will soon provide new information, determining whether the solar flux contains muon or tau neutrinos.

A supernova, the spectacular collapse of a massive star in which the star's outer mantle is ejected, produces prodigious numbers of neutrinos of all three types, which must then propagate through matter billions of times denser than the core of our Sun. The concomitant enhancement of neutrino oscillations in this environment provides another probe of the MSW effect. The experimental challenge is to build neutrino detectors sensitive to all three neutrino flavors and reliable enough to see supernovae, which occur only about once in every 30 years in our galaxy.

The mechanism of supernovae, the creation of the heavy elements, and the nature of nuclear matter at extreme densities all depend on the properties of neutrinos. Supernovae are the main engines driving the chemical evolution of our galaxy, synthesizing new elements and ejecting them into the interstellar medium. Supercomputer simulations of stars so far have been incapable of accurately predicting the synthesis of heavy nuclei in supernovae explosions. Because neutrinos play a crucial role in this process, it may be that new neutrino physics is the missing ingredient in these simulations.

Experiments searching for neutrino oscillations are conducted by producing neutrinos of one type at one location and then detecting these neutrinos at another, distant one. If the type of neutrino at detection differs from the type originally produced, a neutrino oscillation has occurred. The smaller the neutrino mass, the longer this distance must be to see an effect; this is why many experiments use extraterrestrial sources such as the Sun. But experiments in which the neutrino production takes place in the laboratory, under experimental control, provide a flexible alternative to search for neutrino oscillations. For example, tau neutrino masses of the size indicated by Super-Kamiokande might be detectable in long-baseline accelerator oscillation experiments, where neutrinos are created in accelerators and detected in laboratories hundreds of miles away. Both Fermilab and KEK are conducting these important experiments.

Gravity, the Planck Scale, and String Theory

The very short length scale of grand unification is intriguingly close to another important length scale, one involving gravity: the Planck length (see sidebar "Gravity"). Laboratory experiments probe short distances by accelerating particles to high energies. At energies available in today's experiments, gravity is incredibly weak. It is totally negligible in the structure of atoms and nuclei and is important in planetary motion and in stars only because the gravitating masses are so large. But the strength of gravity depends on an object's mass only when it is at rest. Generally the strength of gravity will depend on a particle's total energy, growing stronger with increasing energy. When a particle's energy reaches the Planck energy, the energy necessary to probe physical phenomena at the Planck length scale, gravity becomes as strong as the other forces of nature.

With all the interactions—electroweak, strong, and gravitational—of comparable strength at the Planck energy, it is natural to conclude that this energy is the fundamental scale of the physical world, where all the forces of nature may be unified into a single theory. Einstein's theory, so successful at low energies, is notoriously difficult to interpret once gravity becomes strong, at which point quantum mechanical effects are expected to be important. One exciting development during the past 20 years was the emergence of a framework—string theory—that may lead to a unified theory of gravity with the other fundamental forces in a manner that overcomes these obstacles.

String theory proposes that something profoundly new happens at the Planck scale. A powerful microscope capable of resolving distances on the order of 10^{-33} cm would show a quark or electron to have a physical size. But instead of structure based on still tinier particles, these particles would appear as tiny loops, or strings. String theory is a beautiful theoretical frame-

GRAVITY

Gravity is the weakest of the fundamental forces of nature and unimportant for the physics of atoms, solids, nuclei, and the elementary particles at accessible energy scales. The classical theory of gravity—general relativity—was formulated by Albert Einstein in 1915. In general relativity, gravity is geometry: Mass curves four-dimensional space-time and, in turn, moves in response to its curvature. For much of the 20th century, the progress of experimental gravitational physics was concerned with developing this classical vision in order to understand such large-scale phenomena as black holes, gravitational waves, and the final destiny of stars, pulsars, quasars, x-ray sources, and the universe itself. Some of that progress is described in Chapter 3, "Structure and Evolution of the Universe," which is concerned with these large distance scales.

Yet it was the vision of many scientists that the two great developments of 20th-century physics—quantum theory and Einstein's theory of general relativity—would eventually be unified and that the resulting quantum theory of gravity would be a part of a unified theory of all fundamental interactions. However, the characteristic length scale at which quantum gravity becomes important is the Planck length of 10^{-33} cm or the corresponding Planck energy scale of 10^{19} billion eV. This is a length scale a hundred billion billion times smaller than the size of an atomic nucleus and an energy scale 10 million billion times larger than that of the largest accelerator on Earth. Only in the explosive evaporation of black holes or at the Big Bang, where large and small are one, are such energies realized in this universe. (The upper image at the right shows a numerical calculation of the horizons of two colliding black holes.) But this extreme Planckian scale is important for today's particle physics because it is the scale characterizing the long-sought-after unified theory of all the fundamental interactions described in this chapter.

Remarkably, as described in this chapter, many of the ideas underlying string theory—today's best attempt at the unified theory of all interactions—have their origin in Einstein's 1915 theory of relativity. Curved space-time, dimensions beyond the four of space and time, and black holes all have their place in string theory. (The lower image at the right is a diagram describing the collision of two microscopic strings.) Equally remarkable,

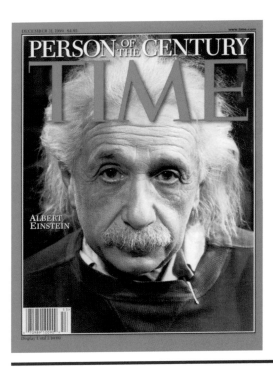

string theory suggests that there is something more fundamental than the space-time that is the focus of Einstein's theory. On this distant frontier, gravitational physics at last becomes important for science at the smallest scales considered by contemporary physics.

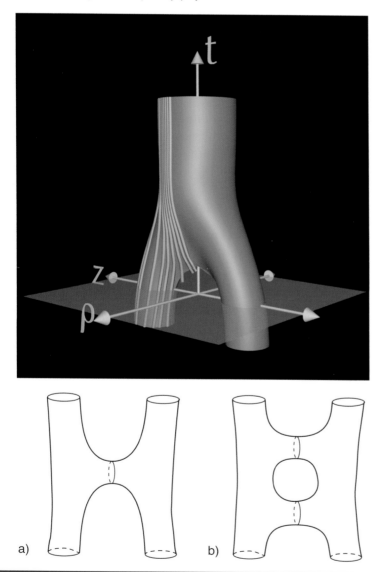

work that is in the midst of intense development. Shaped by compelling ideas of symmetry, including supersymmetry and powerful principles of internal consistency, it incorporates both gravity and quantum mechanics and could answer many of the questions generated by the standard model. If string theory does indeed describe the physical world at the very tiniest distances, its beautiful symmetries are deeply hidden from direct observation. Explaining how this takes place remains an especially challenging problem in theoretical physics.

To some, the current state of development of string theory is reminiscent of the state of quantum theory in the very early years of the 20th century. It was clear then that something radically different from the classical physics of the 19th century was emerging, but it was not until the mid-1920s that the final form of the new theory began to take shape. Although string theory is not directly tied to experiment as quantum theory was during its formation, it does call for radically new ideas, and its future development may prove equally revolutionary.

THE LENGTH SCALES OF NATURE

Our universe has a size, age, and complexity that dwarfs the realm of human experience. Yet physics strives to understand the constituents and forces that shaped all of the cosmos. How can we hope to discover basic laws that simultaneously describe quarks and quasars, transistors and tribology, baseball and black holes?

Our ability to understand such diverse phenomena relies on the same principle affecting the boats described (at p. 10) in the introduction to this volume: Only waves of a certain size, the size of the boat, pose a lethal danger. In a similar way, the phenomena at large length scales are largely insensitive to the detailed laws of nature at shorter scales. The principles that describe earthquakes need not involve the details of nuclear physics; descriptions of Bose-Einstein condensation need not worry about the existence of quarks.

At the shortest distances yet probed, 10^{-16} cm, the standard model of particle physics provides an accurate description of nature. Quarks, the W and Z bosons, gluons, and other exotic particles all must be included to account successfully for the phenomena observed in high-energy accelerators. But when nature is observed at larger length scales, such as the atomic scale, many of these details fade into the background.

A surprising feature of this picture is the remarkable variety of phenomena that arise when we don't try to resolve short distances: For example, the laws that govern atoms are simpler than the standard model, involving fewer particles and forces, yet atoms display extremely complex and often unexpected behavior. Similarly, Newton's laws of motion are simpler than the quantum mechanical equations that describe atoms, yet the movement of Earth's surface in an earthquake is unpredictable. Understanding how these complex properties appear at large distances remains an important challenge.

Although the phenomena at these disparate length scales seem very different, their description is often similar. For example, the description of the hidden symmetry associated with the W and the Z is nearly identical to that of magnetism in some materials. Mathematical techniques developed by string theorists are now being applied to membranes in biological systems. Nothing illustrates the unity of physics better than these similarities across vastly different length scales.

The common ideas of physics have been applied over distances ranging from the realm of string theory to the furthest reaches of our universe. The results have allowed an understanding of a staggering variety of phenomena and lay the foundation for further research as we probe new frontiers at all distances.

Part II

Physics and Society

5

Physics Education

Modern society depends on advanced technology. Decisions ranging from the purchase of an energy-efficient air conditioner to legislation on nuclear power plants now routinely confront the public. Almost all technology is based on scientific principles, and providing people with technical knowledge and scientific literacy in a range of fields is one of the most important missions of the physics community. An education accessible and engaging to a broad spectrum of students at all levels is essential for this mission.

Graduates who have specialized in physics provide a unique component of the technical workforce. With their problem-solving skills and grasp of the principles of physics, they are able to attack a wide variety of problems. A well-trained physicist is capable of moving quickly among different technical areas, particularly into areas so new that they have not yet evolved into an engineering discipline. This ability of physicists to solve problems in a wide variety of fields is illustrated by the number of physicists-by-training who have gone on to win Nobel Prizes in other disciplines—for example, Allan Cormak, Francis Crick, Max Delbruck, Godfrey Hounsfield, and Rosalyn Yalow won in physiology or medicine; Marie Curie, Walter Gilbert, Walter Kohn, Ernest Rutherford, and Alan Heeger won in chemistry; and Andrei Sakharov won for peace.

Physics education must meet the needs of several diverse groups. The general public must have the background they need to understand and foster the progress of science. Industry requires a workforce trained in a wide variety of engineering and science disciplines, all of which are founded on physics principles. And research physicists, scientists who advance knowledge in physics itself, require a lengthy and specialized education.

Meeting these goals has proven to be a difficult task. Enrollment in physics courses declines dramatically as students advance through the educational system, with only about a quarter of students taking a high school physics course. At the university level, physics courses are most often taken as a foundation for study in engineering, medicine, and other sciences and rarely by students in other disciplines. Advanced undergraduate study in physics is undertaken by only a very small number of students, but they are fairly likely to pursue graduate study. In 1999, 31 percent of university and college physics graduates entered graduate programs in physics while an additional 19 percent went into other graduate programs (primarily engineering).

The post-World War II enthusiasm for physics resulted in tremendous interest in the study of physics and expansion of the capacity for physics education in colleges and universities. The space program, which followed the shock of Sputnik and crested with the success of the Moon landing, was accompanied by significant efforts to reform physics education at all levels as well as to expand capacity. The physics community worried that not enough was being done to educate the general public and that its service mission to other academic departments was not being fulfilled. In the early 1980s, the American Association of Physics Teachers, the American Institute of Physics, and the American Physical Society began working together to sponsor programs in support of educational reforms.

An encouraging trend over the past decade was the steady increase in the number of women involved in physics at all levels (Figures 5.1 and 5.2). High school physics courses now have nearly equal male and female enrollments. Unfortunately, the attrition rate at each stage in the educational process up to the Ph.D. level has remained high. The fraction of Ph.D. degrees awarded to women is only 13 percent. It is particularly troubling that physics seems to be lagging other disciplines such as chemistry and mathematics in the elimination of the gender discrepancy.

The representation of most U.S.-citizen ethnic minorities in physics shows a pattern similar to that for women, although the fractions are smaller. There has been a steady increase in the number of minorities receiving degrees in physics, but the overall numbers are small and drop off dramatically the higher up one looks on the educational ladder.

As society becomes ever more dependent on technology, as physics and the other sciences grow closer together, and as physics itself becomes more ambitious and demanding, physics education must change. Beginning with K-12 physics, continuing through college and university physics, and on through graduate and postgraduate education and training, physics education must meet the needs of the general public, the technical workforce, and the research community.

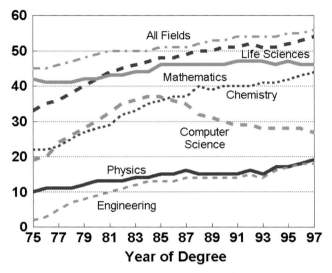

FIGURE 5.1   Percent of bachelor's degrees in selected fields earned by women, 1975 to 1997. SOURCE: R. Ivie and K. Stome. 2000. *Women in Physics, 2000.* AIP Statistical Research Center, College Park, Md.

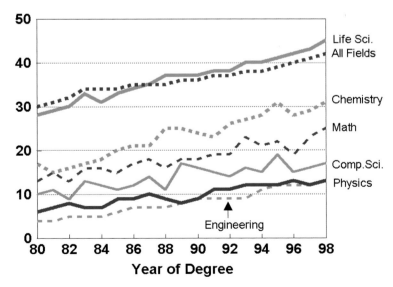

FIGURE 5.2   Percent of Ph.D. degrees in selected fields earned by women, 1980 to 1998. SOURCE: R. Ivie and K. Stome. 2000. *Women in Physics, 2000.* AIP Statistical Research Center, College Park, Md.

K-12 PHYSICS

Good preparation in science at the K-12 level is a prerequisite to good science education at all levels. Physics education begins as part of general science in grade schools and becomes increasingly specialized as the grade levels increase. Typically, a course labeled "physics" has been an option taken late in high school. During the 1970s and 1980s, there was a significant decline in the percentage of students who were studying physics in high schools. In the past decade that decline has reversed itself (see Figure 5.3).

Several studies have documented the problems with education in the physical sciences at the K-12 level. One is the Third International Mathematics and Science Study (TIMSS), which carried out detailed comparisons between the ways students are taught mathematics and science in different

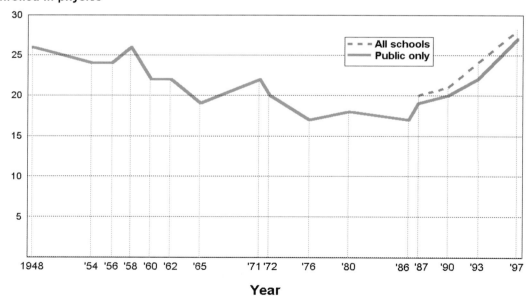

FIGURE 5.3   Percentage of high school seniors enrolled in physics, 1948 to 1997. SOURCES: AIP High School Physics Teachers Surveys (1986-1987, 1989-1990, 1992-1993, and 1996-1997); AIP (1964); G. Pallrand and P. Lindenfield, 1985, "The Physics Classroom Revisited: Have We Learned Our Lesson?" *Physics Today* (November) pp. 46-52; Department of Education, National Center for Education Statistics (various years).

countries and how they perform on a number of exams. Students in the United States are particularly weak in their understanding of physical sciences at all grade levels, and this is dramatically worse at the higher grades. U.S. students who have had a high school physics course score lower on tests of understanding of physics than comparable students of any other country evaluated. TIMSS characterized U.S. teaching as primarily "learning terms (definitions) and practicing procedures." Conspicuously absent is the teaching of the nature of the scientific process and the concepts of physics as they apply to the world around us. These are the elements of science education most important for a technically literate society and workforce.

Results such as those from TIMMS have led to concern about the preparation of K-12 science teachers, in particular their background in the discipline they are teaching. Only about a third of the U.S. physics teachers majored in either physics (22 percent) or physics education (11 percent), with many of the others having majored in another science or mathematics (Figure 5.4). This is a significant problem, because a lack of knowledge of the material is frequently cited as a contributor to poor teaching. One reason for the relatively weak training in physics of K-12 teachers is probably the minor role that physics departments now play in the training. The

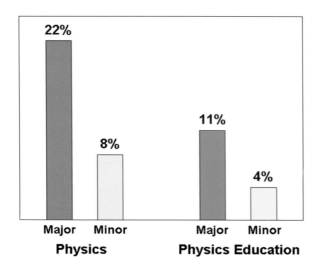

FIGURE 5.4   Percent of high school physics teachers with physics degrees, 1996-1997. SOURCE: AIP High School Physics Teachers Survey, 1996-1997.

responsibility for this training is largely in the hands of the education schools, and with rare exceptions they have little connection with the physics departments. There is a growing sentiment that teachers can be better prepared by giving physicists a greater role in the process. Implementing this approach will require the encouragement and cooperation of university administrations and education schools as well as changes within the physics departments. The importance of these changes was recognized in resolutions adopted independently by the American Institute of Physics, the American Physical Society, and the American Association of Physics Teachers. The resolutions encouraged physics departments to take increased responsibility for the education of teachers.

In addition to taking a greater role in preparing future teachers, physics departments can contribute to K-12 education by providing ongoing training and assistance to current K-12 teachers. Although many isolated local efforts do this, there is a need for broader programs. The mathematics community has convinced the National Science Foundation (NSF) to fund a faculty member in selected mathematics departments whose main role is to interface with the high school teacher and student community. This is one possible model. Another is the establishment of programs such as the Physics Teaching Resource Agents (PTRA). A cadre of highly qualified and highly trained high school teachers, the PTRA worked with colleagues from higher education to reach out to the majority of those teaching physics who did not have a background in physics. Prominent physicists from our universities and national laboratories have also become involved in local programs reaching out to high school teachers and students. Programs that provide research experiences for K-12 teachers, such as NSF's Research Experiences for Teachers program, can be very effective. They can convey the excitement of real research to teachers and, through them, to even larger numbers of students. All these programs require adequate release time for teachers to take full advantage of them.

UNDERGRADUATE PHYSICS

In the scientific community and in the leadership of many universities, there is growing concern about the decreasing numbers of students majoring in science. Seymour and Hewitt[1] have studied the reasons for the high

[1]E. Seymour and N. Hewitt. 1997. *Talking About Leaving: Why Undergraduates Leave the Sciences*. Westview Press, Boulder, Colo.

attrition rates for students who enter college intending to major in science. The attrition rate of greater than 60 percent for physical sciences is considerably higher than for any other major (Figure 5.5). They find no evidence that the students who switch to nonscience majors are less capable than those who do not switch or that inadequate high school preparation is a more serious concern for students who switch than for those who continue. However, a factor that does cause students to switch to other majors is the perception that these alternative majors provide a better education and are more intrinsically interesting.

It is notable that the factors causing women and ethnic minorities to switch out of pursuing science careers are similar to those motivating all students to switch, but the effect of these factors, and hence the attrition rates, seems to be magnified for these underrepresented groups.

Physics is often singled out for particular criticism. Sheila Tobias documented how introductory physics courses deterred many who took them

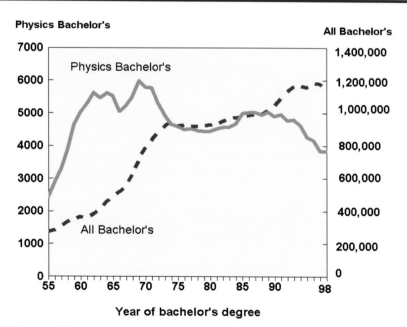

FIGURE 5.5   Physics bachelor's and total bachelor's degrees produced in the United States, 1955 to 1998. SOURCES: American Institute of Physics Statistics Division, *Enrollments and Degrees Report*, 1999, and National Center for Education Statistics, *Digest of Education Statistics*, 1999.

and did not continue in physics.[2] She selected her sample population from those who became successful in other areas. She also recruited highly educated volunteers to take introductory physics courses and to report on their experiences. She paints a compelling picture of introductory courses that do not encourage the further study of physics.

Introductory Curricula

Excellent introductory physics courses are important for all students, from those planning to become professional physicists to those majoring in the arts and humanities. Physics also plays a foundational role in the other fields of natural science. Even where the full rigor of physics is not applicable, the approach of physics to characterizing and analyzing problems can be very useful. Every field of natural science and engineering has made use of the instrumentation and experimental techniques developed by physicists; these techniques are an essential part of a physics education.

Over the past two decades, the American Physical Society, the American Institute of Physics, and the American Association of Physics Teachers, among others, devoted much attention to problems in introductory physics courses. Research in physics education began to demonstrate that students were not learning physics as well as had been hoped or expected. Many of the reports and projects began to call for revisions of the curriculum that would do the following:

- Take account of research findings on physics education.
- Introduce modern, topical physics earlier in the undergraduate curriculum. Make connections to other areas of science.
- Focus more on concepts and discovery rather than covering a prescribed set of topics.

The research underlying many of the reform efforts called for more interactive forms of instruction ("active engagement" was a term used). The approach gained momentum after the introduction of a test of the conceptual understanding of mechanics that became known as the Force Concept Inventory. This test and its successors allowed faculty to compare innova-

---

[2]Sheila Tobias. 1990. *They're Not Dumb, They're Different: Stalking the Second Tier*. Research Corp., Tucson, Ariz.

tions through a system of pre- and post-testing of students that revealed the differential learning gains.

The results of these tests of physics learning often surprised physics faculty, who had assumed that students were learning far more than they actually did. In the various interactive learning formats, the instructors often became aware of student difficulties that they had not observed in the less interactive lecture format. It was important to be able to document student performance in the traditional lecture courses so that student difficulties would not be (erroneously) attributed to the interactive formats. The existence of a body of data covering students at quality institutions all across the country helped to set a baseline for evaluating the new programs.

Introductory physics courses at the universities come in several varieties. One way to classify them is by the level of mathematical sophistication involved. Large universities often offer a spectrum of courses that can be characterized as follows: conceptual physics (no mathematics or minimal mathematics), college physics (trigonometry-based), and calculus-based university physics. Introductory courses are sometimes tailored to the needs of different majors, with the predominant course of that kind being the engineering physics course. Other majors that are often addressed separately include the medical and life sciences, architecture, and the liberal arts. The diversity of offerings belies the similarity in content, coverage, and approach that characterized introductory physics education throughout the last century and up through the present. The texts for these introductory courses all tend to have much the same logical development and subject coverage and changed little over the course of the 20th century.

In 1987, the American Institute of Physics launched the Introductory University Physics Project (IUPP) to recommend alternative approaches. The IUPP focused on content, coverage, and order rather than on pedagogical approaches and alternative methods of organizing instruction. Efforts to include modern physics topics such as quantum mechanics, particle physics, and condensed matter physics were an important part of this project.

Over the past few years, a number of curricula have been developed in the United States using the research/curriculum reform/instruction cycle. Classes based on this approach ("active engagement" classes) have in common a focus on what the students actually do and what the effect of that activity is. An excellent example is Physics by Inquiry, a model developed by the University of Washington that has set a standard for others to aspire to. This discovery-learning, research-based curriculum has gone through 20 years of development and testing. It guides students through the reasoning and construction of scientific ideas through hands-on laboratory experience. A few generic models for active engagement classes follow, along with specific examples of courses:

- Studio/workshop models
  - Physics by Inquiry
  - Workshop Physics (Dickinson College)
  - The Physics Studio (Rensselaer Polytechnic Institute)
- Discovery labs
  - Tools for Scientific Thinking (University of Oregon)
  - RealTime Physics (University of Oregon and Dickinson College)
- Lecture-based models
  - Active Learning Physics System (Ohio State University)
  - Peer Instruction/ConcepTests (Harvard University)
  - Interactive Demos (University of Oregon)
- Recitation-based models
  - Cooperative Problem Solving (University of Minnesota)
  - Tutorials in Introductory Physics (University of Washington)
  - Mathematical Tutorials (University of Maryland).

Departments can begin to draw on this experience by determining what they want their courses to accomplish and critically evaluating whether or not they are achieving it. As part of this evaluation they could benchmark their introductory programs against the leading innovative programs and revise their offerings as appropriate. Questions that need to be answered include the following:

- Are students actively engaged in doing physics rather than watching the instructor?
- Do students in introductory courses gain an appreciation for physics as it is done today?
- Does the introductory course take advantage of the computing and communication tools available today?
- Do students leave the class with a sense of the excitement of physics, its connection to the other sciences, and how it applies to the world around them? Do they consider further study?

The committee believes that high-quality introductory courses are essential for creating a 21st-century workforce, for fostering scientific literacy, and for attracting and retaining talented American students to further study in physics.

Advanced Undergraduate Programs

With the trend toward incorporating more "modern" physics in the introductory courses and with the increasing sophistication of students in the use of technology, there is an opportunity for physics departments to renew their advanced undergraduate courses. It is important for students to see how physics has contributed to the biomedical sciences and to the growth of information technology. Physics departments would do well to collaborate with their colleagues in other departments to encourage cross-disciplinary experiences and to implement minors and double majors that combine physics and biology, physics and computer science, physics and biomedical engineering, and other novel combinations.

The revolution in information technology and biotechnology has made thousands of scientists into entrepreneurs. For example, Jeff Kodosky, a physics major, cofounded National Instruments based on his vision for how physicists (and other scientists) should be able to interact with scientific instrumentation. Physics departments can work with schools of management to provide physics students with role models and business models for using physics to create new enterprises and new value within existing enterprises (see sidebar "Research Experiences for Undergraduates").

In physics, teaching and learning flow from research. It is difficult to imagine a proper learning environment for physics students that does not give them extensive exposure to research and physicists engaged in research. These opportunities can be found at major facilities, research centers of various kinds, and research projects under the direction of single investigators. This vision for undergraduate research experience can be more difficult to achieve in smaller liberal arts colleges, but there are indeed opportunities and outstanding examples. The Council for Undergraduate Research was formed specifically to encourage this activity at schools of all sizes. Research universities can, do, and should offer summer and other short-term research experiences to undergraduates from all institutions. Support for this activity can often be built in as supplements to research grants or as targeted funding programs from the NSF.

It can be a challenge to create experiences that benefit both the research effort and the student. Undergraduate students may not have the knowledge to contribute to some aspects of advanced research projects, and it may be difficult to identify tasks that would be meaningful for the project and the student, but physics departments should nonetheless assign a high priority to supporting and integrating students into research projects.

RESEARCH EXPERIENCES FOR UNDERGRADUATES

The most important factor governing the future health of the nation's physics enterprise is the quality of the young people drawn to the field. The undergraduate years are a crucial time in the career paths of potential scientists: These years present the first opportunities for in-depth study that may influence career decisions, convincing potential scientists that the excitement of research and discovery is ample reward for the hard work that science requires.

While there are many community efforts to enhance opportunities for undergraduate research, one of the most successful programs is that begun by the National Science Foundation a decade ago. The NSF Research Experiences for Undergraduates (REU) Program helps interested sites—university departments, government and industrial laboratories, and other research organizations—to recruit undergraduates to take part in individual research projects under the guidance of interested faculty and other senior researchers. The goals also include broadening the talent base in physics and other sciences: Many REU programs have succeeded in increasing participation by women and members of underrepresented minority groups. REU research experiences can be particularly valuable for students coming from smaller colleges that lack the laboratory facilities of larger institutions.

Several thousand students now participate in REU programs in mathematics, science, and engineering. The response by physicists has been particularly strong: More than 100 sites have been created, providing 1000 undergraduate research positions each year. This represents an extraordinary effort by the site directors and the research advisors who give their time to these programs. Most of the sites operate during the summer and provide stipends and housing support for aspiring scientists willing to trade their summer vacations for an opportunity to do research. Below: REU students at LIGO, Hanford, Washington.

GRADUATE EDUCATION

Graduate and postdoctoral physics education programs in the United States are the envy of the world. They are the central sources of high-quality scientific personnel needed to sustain the high levels of achievement that have characterized U.S. physics in all its venues since the end of World War II. Yet physics is changing, and the source and destination of students in the nation's graduate programs are changing also. These changes require corresponding changes in the graduate and postdoctoral education programs, changes that the committee describes below.

Supply

The number of U.S.-educated undergraduates in physics has decreased in the last 15 years, leading to an imbalance between the supply of U.S. bachelor's degree holders and the capacity of U.S. graduate programs (Figures 5.6 and 5.7). Physics departments have reacted by increasing the flow of students from other countries. The committee considers this is a healthy development in view of the anticipated demand. Advantage should be taken of the best talent wherever it may be found. In any case, physics in the United States has always been enriched by this flow of talent from abroad. The flows of physicists from Europe before and immediately after World

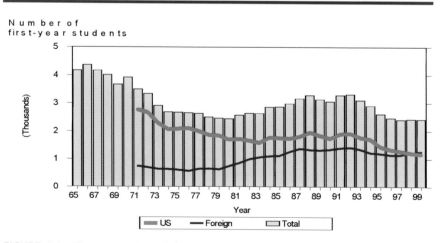

FIGURE 5.6   First-year U.S. and foreign physics graduate students, 1965 to 1999. SOURCE: American Institute of Physics Statistics Division, *Enrollments and Degrees Report.*

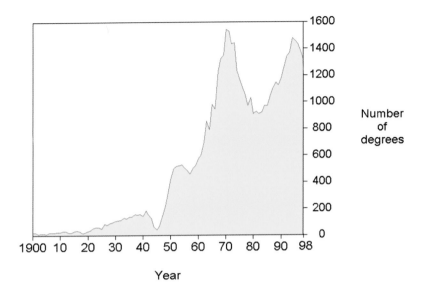

FIGURE 5.7   Number of physics Ph.D.'s conferred in the United States, 1900 to 1998. SOURCE: American Council on Education (1900-1919), NAS (1920-1961), American Institute of Physics (1962-1998), American Institute of Physics Statistics Division, *Enrollments and Degrees Report.*

War II, from eastern Europe and the former Soviet Union after the collapse of the latter, and from Asia during the last decade's political unrest in China have all brought highly talented individuals to this country. While celebrating the flow of talent from other countries, the committee also expresses its concern about the significant decrease in the number of U.S. students entering graduate study in physics, reflecting the drop in the number of bachelor's degrees awarded in physics.

Destination

The production of Ph.D.'s in physics has been highly volatile over the past 30 years, as has been the demand for them. However, one trend can be clearly identified: An increasing fraction of physics Ph.D.'s find employment in industry. Of the graduates in 1996, 47 percent entered industry while 41 percent entered academia.

Implications

Most graduate programs have not changed much from the 1960s and 1970s, when the majority of Ph.D.'s were going into academic teaching and research positions. In the committee's view, the structure of graduate programs should better reflect the changing and varied destinations of students who pursue advanced degrees. Physics departments need to reassess the missions of their graduate programs in light of where the students are headed.

As physics becomes increasingly connected with other sciences and more important in technology development, the committee foresees a growing need for programs that provide training for the majority of students who will not pursue careers in academia. The increasing employment of physicists in industry and the appreciable crossover of physics graduates into other fields support this conclusion.

There is currently a great need for workers with technical expertise, and a graduate physics education is well suited to provide that expertise. Strong physics training provides students with broad problem-solving skills and familiarizes them with a wide range of technologies and the underlying physical principles. This allows them to adapt easily and contribute to many different areas such as electronics, optics, and computational modeling.

Many technical careers do not require training to the degree of specialization or for the extended time (typically 6 years) required to obtain a physics Ph.D. A master's degree would appear to be a more appropriate level of physics education for students interested in working in many high-tech industries. Such training would also be valuable for students planning to work in nonphysics areas that require a good knowledge of physics, such as atmospheric science, radiology, and many types of advanced engineering. While there has been growth in professional master's degrees in areas such as engineering, computer science, and management, physics has not exploited this educational model significantly. Some other opportunities for creating professional master's degrees in physics include the following:

- Physics for those going into information-technology-related areas,
- Physics for incipient entrepreneurs,
- Physics and intellectual property law, and
- Physics and the continuing education of engineers and others.

SUMMARY

Physics education at all levels must focus on producing a scientifically literate public and a technically trained workforce. High-quality physics courses are an essential component of science education. They are one of the best avenues to providing the public with the knowledge to make informed scientific decisions, as well as providing technical training for the modern workforce. In addition, these courses are crucial for attracting and retaining capable American students for further study in physics.

Introductory course offerings need to be improved. Educators should take note of innovative teaching methods that have been shown to be effective. Introductory courses are the last exposure to physics that most university graduates will have, so they must lay the physics foundation for further technical training in all areas. For those continuing in physics, the courses must engage students, decrease attrition, and attract new recruits. Courses should be revised so they introduce students to concepts and questions of modern physics, make increased use of advanced computing and communication technologies, and incorporate active physics engagement techniques.

Advanced undergraduate and graduate curricula should reflect physics as it is currently practiced, making appropriate connections to other areas of science, to engineering, and to schools of management. High-quality undergraduate research opportunities are an important tool for introducing students to modern physics practice. Physics education needs to reflect the career destinations of today's students. Only a third of all physics majors pursue graduate degrees in physics, and of those who do, nearly three-quarters will find permanent employment in industry. The undergraduate and graduate curricula must satisfy the educational needs of these students.

The education of K-12 teachers benefits greatly from the involvement of professional physicists. Physics departments can become more involved in the training of high school teachers by offering courses that are geared to the education of future physics teachers and by creating and conducting outreach programs. This conclusion is supported by the broader physics community through the resolutions of the American Physical Society, the American Association of Physics Teachers, and the American Institute of Physics. Achieving each of these goals will be difficult, requiring changes in university physics departments, encouragement from university administrations, and support from state and local education boards.

6

Health and Biomedical Sciences

Like all aspects of the modern world, medicine has been transformed by discoveries in physics during the past century. Modern medicine is unthinkable without modern chemistry, and modern chemistry could never have existed without the fundamental understanding of atoms and molecules provided by the revolution in physics that started just about 100 years ago. A series of advances in physics itself has also had a decisive and direct impact on medical science as we know it today, starting with the discovery of x rays, for which Roentgen was awarded the Nobel Prize in 1901, and the discovery of radioactivity, for which Becquerel and the Curies received the Nobel Prize in 1903. Today nearly all the therapeutic and diagnostic tools of a modern hospital have their origin in basic physics research.

THERAPY

Ionizing radiation remains one of the three options for treating cancer. Either alone, or in combination with surgery and chemotherapy, ionizing radiation is used as therapy for most malignancies. Over the years, the use of ionizing radiation has become more sophisticated, and it is highly effective for certain types of tumors. Recently, for example, the radiation transport absorption coefficients and results from the simulation programs used by nuclear weapons designers have been applied to the problem of optimizing radiation dose in cancer treatment. The result is a new dose calculation system that for the first time can model the varying materials and densities in the body in combination with the radiation beam delivery system.

The breakthrough brings to the field of radiotherapy a new level of accuracy in the ability to predict where radiation deposits energy in the body, opening the way to more accurate prescriptions for radiation therapy, more aggressive treatment of tumors, and substantially lower risk to normal tissue. For example, this new approach will allow more aggressive treatment of prostate cancers without endangering the bladder and spinal tissue adjacent to the prostate gland.

Lasers are a much more recent invention, but their therapeutic use is rapidly increasing. Focal tissue coagulation by laser is the standard nonsurgical treatment for a detached retina, and the reforming of the corneal surface shape (laser keratotomy) with laser radiation is a very effective treatment for nearsightedness that is now becoming standard. The use of local laser heating is a microsurgical technique that is finding application in a variety of fields.

Although of more limited use, ultrasound has also found an important place as a nonsurgical treatment for kidney stones and for the cleaning of surgical instruments.

Another revolution in therapy, based on the use of fiber optics, has been developing over the past several decades. Before physicists developed ordered fiber-optic bundles, visualization of internal body surfaces was limited by the requirement that light must follow straight lines; this meant, for instance, that only small portions of the gastrointestinal tract and the airways could be seen by a physician. With fiber-optics and video imaging, the gastroenterologist and otolaryngologist can now see much of the inner surface of humans. A more recent trend has been to use fiber-optics imaging to permit surgery through tiny incisions. Many appendectomies and other surgical procedures—even including heart surgery—are now carried out in this much less invasive way. Because damage caused by large incisions and the exposure of internal organs is minimized, these remote surgical procedures can be less dangerous and patients can recover more quickly and with less discomfort.

New materials have also become available to replace damaged or missing body components. Silicone is used in reconstructive surgery to provide internal support for soft tissues, and new, hard plastic materials are transforming dentistry by permitting better reconstruction of damaged teeth. These materials not only are strong but also bond well to the tooth surface and expand with heat like the natural tooth; this means that a much wider range of tooth damage can be repaired, and that the repairs are much more durable than in the past.

DIAGNOSIS

The use of ionizing radiation has been important therapeutically for more than half a century, but physics also plays an ever more important role in diagnosis. Because such small quantities of radioactive material can be detected, radioactive labels have become very important in a variety of diagnostic contexts. Labeling red blood cells with radioactive isotopes of chromium permits the lifetime of these cells to be measured, and this method can be used to determine if anemia is the result of decreased production or increased destruction of the red blood cells. The radioimmune assay—Yalow received the Nobel Prize in 1977 for the development of this technique—makes use of antibodies that have been made radioactive; these antibodies detect minute quantities of hormones and other chemicals by binding to them and providing a radioactive tag that can be used to detect the presence of the molecule. Because signaling molecules, like hormones, are present at minute concentrations in the blood and other bodily fluids, other methods for detecting their presence are too insensitive. Some organs specifically take up certain atoms or chemical compounds (the uptake of iodine by the thyroid gland is the best-known example), and this fact has enabled physicians to assess organ function or identify the presence of damage by monitoring the uptake of these substances that have been tagged with radioactivity.

Probably the most striking advance in medical diagnostics has been the development of remarkable imaging techniques. Our ability to look inside the living body noninvasively started with Roentgen's discovery of x rays about a century ago. X-ray machines soon became common and were the standard—and only—method for diagnostic imaging until the recent explosion in imaging technology. As powerful computers became available, the data contained in images could be manipulated to extract the available information about three-dimensional structures present in two-dimensional images made from different views. Cormak received the 1979 Nobel Prize for developing computerized tomography, the now-standard method for extracting such three-dimensional information from two-dimensional projections.

Versions of this tomographic method have been extended to other types of imaging. Magnetic resonance imaging (MRI), formerly called nuclear magnetic resonance imaging, is based on the discovery of magnetic resonance in nuclei, for which Rabi received the 1944 Nobel Prize in physics. Rare only a few years ago, MRI is now standard and practiced in virtually all medical centers (see sidebar "Computerized Tomography"). Because of the remarkable resolution of this method—brain structures in the millimeter size range are clearly imaged—MRI and its more recent modifications, such as magnetic resonance angiography (for viewing blood vessels in the heart), have transformed many areas of medical practice, from neurology to the surgical specialties.

COMPUTERIZED TOMOGRAPHY

The modern descendants of the venerable x-ray machine use computer technology to greatly improve the quality of the images produced. Computerized tomography (CT) takes information from two or more slightly different views of the same object and reconstructs a three-dimensional image of the object that can be rotated and sliced by computer. This technique is now applied to x-ray images and also to novel imaging methods such as MRI (magnetic resonance imaging) and SPECT (single-photon-emission computerized tomography) imaging. On the top is a pair of slices through an image of a head made by MRI, and below a similar pair of slices of an image of the same head produced using SPECT. Atoms such as hydrogen (in water) have magnetic properties, and these subatomic magnets can be aligned by a very strong magnetic field; how fast the tiny magnets line up depends on their surroundings, and this is exploited by MRI to produce images of tissue in the living body. Certain radioactive agents that have been injected into the bloodstream, like technetium-99, emit photons, and so brightness of the "light" emitted is used by SPECT to tell how much of the agent is present and therefore how much blood is flowing to that region. Because organs increase blood flow according to their needs—the brain does this for very small regions devoted to particular functions, such as language—the colors that appear provide a picture of the brain's activity at each location. Both of the methods that

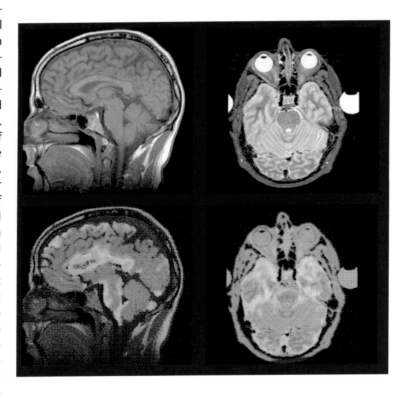

gave these images have become standard ways for evaluating organ structure and function, illustrated here by images of our most complex organ, in health and disease.

Writing recently in the *Washington Post*, former NIH director Harold Varmus pointed to MRI to underscore the important contributions of the physical sciences to health care:

> Medical advances may seem like wizardry. But pull back the curtain, and sitting at the lever is a high energy physicist, a combinational chemist or an engineer. Magnetic resonance imaging is an excellent example. Perhaps the last century's greatest advance in diagnosis, MRI is the product of atomic, nuclear and high-energy physics, quantum chemistry, computer science, cryogenics, solid state physics and applied medicine."[1]

Related techniques developed by the physics community, such as functional magnetic resonance imaging (fMRI) (see sidebar "Functional Magnetic Resonance Imaging") and positron emission tomography (PET), are now being applied to understand the functioning of the human brain. Researchers are using these high-resolution scanning techniques, which are sensitive to local brain metabolism or even to local concentration of complex psychoactive biochemical molecules that are thought to regulate brain function, to diagnose and guide treatment for a number of mental illnesses and for some forms of drug addiction.

For example, brain scans have identified depressed metabolic rates and regions of abnormal biochemical concentrations in those regions of the brain associated with impulse control. Researchers have correlated these deficiencies with patients who have been diagnosed with attention deficit disorder (ADD). This opens the possibility of both a direct diagnosis of this condition and a way to monitor the effect of medications such as Ritalin on the brain function of patients with ADD.

Similarly, brain scans are being used to identify local metabolic or biochemical anomalies in patients with Alzheimer's disease and with addictions to cocaine and other substances. These scans also offer the possibility of directly identifying the underlying disorder and a way of monitoring directly the impact of medications and treatment.

The revolution in imaging made possible by the availability of powerful computers extends to the use of ultrasound. In many areas of medical practice, particularly obstetrics, ultrasound images are standard and can be used to follow changes over time, such as fetal growth, with a resolution sufficient to detect structural abnormalities or determine fetal sex.

Advances in imaging also extend to light microscopy. Zernike's invention of phase-contrast microscopy (for which he received the Nobel Prize in

---

[1]Harold Varmus. 2000. "Squeeze on Science," *Washington Post*, October 4, p. A33.

FUNCTIONAL MAGNETIC RESONANCE IMAGING

Functional magnetic resonance imaging (fMRI) works by mapping tiny magnetic disturbances caused by changes in the amount of oxygen present in tissue. Since the brain very precisely regulates the blood supply to support the high metabolic demands of active neurons, changes in fMRI signals reflect changes in neural activity. In the image below, fMRI-measured neural activity is projected onto a three-dimensional rendering of a subject's brain. For visualization purposes, the cortex has been computationally inflated so the depths of the brain's folds are revealed; the view is slightly from the back so that some of the right side can be seen. The boundaries of the various cortical regions that process visual information (V1, V2, V3A, MT) are defined by experiments that map the cortical representation of the visual world. The colored pixels within the visual area boundaries represent the results of an experiment measuring the effects of visual spatial attention. During the experiment, a subject had to detect a stimulus that appeared on either the left or the right side of the visual field. The subject was told to attend alternately to the left and the right, but neither the stimulus nor the subject's eye position changed. In the image, the red pixels represent activity modulated during the attend-right phase of the experiment, and the green pixels, activity modulated during the attend-left phase. The red region labeled MT is the cortical area specialized for detecting movement.

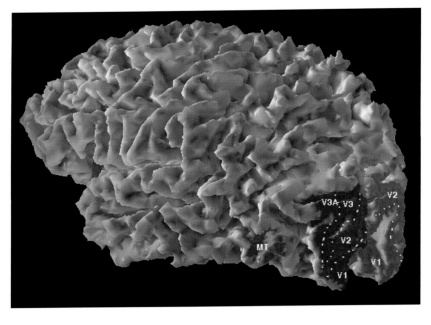

1953) was the first in a series of improvements and extensions of traditional light microscopy. Differential interference contrast, videomicroscopy with computerized manipulation of the image, infrared microscopy, laser scanning microscopy, and two-photon microscopy are recently developed microscopy techniques that have become standard for research and are beginning to be used for medical diagnosis.

UNDERSTANDING THE BODY

The basis for medical practice resides in an understanding of how the body functions. Physics has contributed to this knowledge in essential ways through biophysical research and through the development of enabling technologies for the biological sciences. At the level of molecules, structure and function are inseparable. Proteins are giant molecules that provide cells with their internal mechanical support and are responsible for carrying out or enabling all of the chemical processes at the core of life. Over the past several decades, biophysicists have determined the exact atomic structure of hundreds of proteins by x-ray crystallography (the technique for which the Braggs won the Nobel Prize). More recently, synchrotron radiation, produced at several of the large particle accelerator facilities (Lawrence received the Nobel Prize in physics in 1939 for inventing the first of these accelerators, the cyclotron), has been used increasingly in place of traditional x-ray machines to determine protein structures more rapidly and accurately. Because of this vast amount of work, we now understand just how many proteins carry out their jobs.

Understanding the structure of complex biological molecules is also the key to developing new pharmaceuticals by an approach that is more rational than the hit-and-miss approach of the past. Viruses are an important example. A virus enters a cell by penetrating the cell membrane after docking to a surface protein, a process that can occur only when the shape of the virus fits the cell's protein as a key would fit a lock. Determining the underlying structure of a virus such as the common cold virus or HIV opens the possibility of developing pharmaceutical molecules that compete with these viruses for the docking sites and thus prevent viral penetration into the cell.

Most drugs work by binding to unique sites on specific proteins and modifying how the protein does its job. The search for molecules that can bind selectively to proteins and alter their function is very difficult and inefficient if done by trial and error. It is the application of physics techniques such as x-ray crystallography that allows determining the structure of viruses and identifying candidate sites on a virus to mimic with a therapeutic molecule. This rational drug design is now a major aspect of research at the many synchrotron radiation facilities that have been developed by the physics community in the past decade. A number of large pharmaceutical

OPTICAL TWEEZERS

Novel techniques, most of them invented in the last several decades, are being used increasingly to study individual molecules of importance to biology. A good example is measuring the "horsepower" of kinesin molecules, illustrated in the figure. Kinesin is a molecular motor, one of dozens of different sorts of specialized protein motors that are responsible for all forms of biological movement in all living organisms, from bacteria to humans. The most familiar manifestation of molecular motors at work is our ability to generate force with our muscles, but the beating of our heart, the tension of blood vessels that sets our blood pressure, the division of one cell into two during growth, and the movement of molecular cargoes from one place in a cell to another are all jobs that depend on a variety of molecular motors. Kinesin is a motor that moves along tiny "railroad tracks" called microtubules to carry important molecular structures from one place in a cell to another.

The strength of single motors was recently measured at Stanford University. Physicists there used optical "tweezers," a method that generates a force on small objects by laser light focused at a tiny point. The kinesin motor was attached to a small bead and then permitted to "walk" along its track, a microtubule. The force generated by the kinesin could be measured by seeing how hard the tweezers had to pull back on the bead to keep the motor from moving along the microtubule. Experiments of this sort are revealing how chemical energy, ultimately derived from the food we eat, is turned into forces that we use to affect our environment, both external and internal.

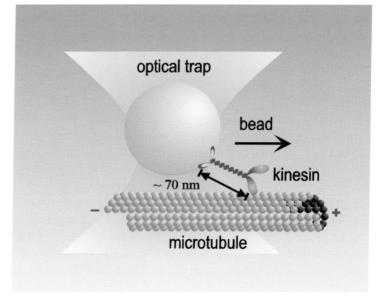

companies and the NIH have invested many million of dollars in instrumentation at these facilities to support this kind of research.

Knowing the structure of proteins is essential for understanding their function, but we must also learn how parts of proteins move as they carry out their tasks. Biophysicists have combined the methods of modern molecular biology with sophisticated physical techniques to learn just how many types of proteins do their job. All cells contain tiny molecular motors—these motors are a large family of proteins—that help the cell to move or that move material within the cell from one place to another. The most familiar manifestation of these motors at work is muscle contraction and the pumping of the heart, but molecular motors are essential for the jobs performed by all cells, from the dividing of cells to produce replacements for cells that die to the carting of essential chemicals to where they are needed in the brain (see sidebar "Optical Tweezers"). Unraveling how these motors work has required developing clever physical methods to measure the motions of single protein molecules over distances that are smaller than can be seen with microscopes.

The beating of the heart and all functions of the brain require another kind of protein that produces electrical signals in cells. The electrocardiogram and the electroencephalogram are manifestations of the operation of these proteins and are used for the diagnosis of heart and neurological diseases. Biophysicists have also learned how these proteins generate their electrical signals, and this knowledge has given us many therapies, ranging from drugs to treat abnormal heartbeats to the treatment of epilepsy.

SUMMARY

In small ways and large, from the treatment of nearsightedness to the diagnosis of neurological diseases, applications of the discoveries of physics have revolutionized medical practice. Although this transformation began 100 years ago, it has greatly accelerated in the past several decades as our increased ability to process information has been applied to a variety of physical phenomena. In the 21st century, physics will continue to provide the foundation for striking advances in biomedicine and health. The increasingly interdisciplinary nature of physics research is focusing more of the field's intellectual and technical resources on addressing opportunities in biology and medicine, both to increase our understanding of basic mechanisms and to provide novel technologies for research and application. In the coming decades we can confidently expect that physics will continue to contribute to the nation's health and, in fact, that these contributions will increase.

7

The Environment

Physics lies at the core of the earth sciences. It is essential for understanding the deep structure of Earth and the natural phenomena that affect Earth's surface, such as earthquakes and volcanic eruptions. These topics, along with others aspects of the physics of Earth, are discussed in Chapter 2, "Complex Systems."

Physics also provides a basis for understanding the dynamic interactions between the atmosphere and the oceans and for the study of short-term weather and long-term climate change. This understanding is essential to stewardship of the environment: for addressing problems like urban air pollution and lake acidification and for dealing with natural hazards such as floods and hurricanes.

Much of physics is the study of energy and its transformation, and energy lies at the heart of important environmental issues. Climate is shaped by how the energy of the Sun affects movement of the atmosphere and oceans and how they in turn distribute energy around the world. Most of the impact of humans on the environment revolves around the need for energy production.

To understand the complexities of the environment and to address problems effectively, the underlying physics must be combined with chemistry, geology, atmospheric and oceanic science, and biology. The ocean-atmosphere system, environmental monitoring and improvement, and energy production and the environment are three areas where an understanding of the basic physics has played a central role and where it is crucial for further progress.

THE OCEAN-ATMOSPHERE SYSTEM

In the years ahead, a continuing improvement in our understanding of the remarkable concentration of energy involved in severe weather patterns

116

will lead to much greater predictive capability and earlier warning than are available today. This progress will come from a combination of theoretical modeling, computer simulation, and direct measurement, each drawing on the tools of physics and each conducted by researchers schooled in the methods of physics.

Until the 1980s, atmospheric science had concentrated on the theory and practice of weather forecasting, which involved a time scale of 6 to 10 days. Weather forecasting was based on an understanding of the prevailing instability of large-scale, mid-latitude phenomena resulting from an analysis of the Navier-Stokes fluid dynamic equations. In the oceans, meanwhile, the emphasis was on attempting to understand the physical processes that accounted for mass and heat transport in cases such as the Gulf Stream and the circulations of the ocean basins. At that stage of understanding, it was thought that variability in the oceans and variability in the atmosphere were relatively independent of each other on time scales shorter than decades.

More recently, it has been realized that the ocean and the atmosphere are coupled on much shorter time scales. This realization emanated from the developing understanding of the El Niño phenomenon in the Pacific Ocean. A series of positive and negative feedbacks between the ocean and the atmosphere create this phenomenon, an oscillation on a grand scale, which is responsible for an instability of the climate system in the Pacific region. The understanding of this phenomenon, which rests on the joint fluid dynamics of the ocean and the atmosphere, suggests a predictability in the climate system. Predictability has been demonstrated not only on the weather time scale of 6 to 10 days but also on an interannual time scale of 6 months to 1 or 2 years, the time scale of the El Niño-coupled ocean-atmosphere instability. Since the pioneering work on the El Niño phenomenon, it has been shown that the great monsoon systems of the planet are also coupled ocean-atmosphere phenomena on the same time scale, so that their evolution depends on the same joint dynamics and thermodynamics of the atmosphere and ocean.

ENVIRONMENTAL MONITORING AND IMPROVEMENT

An ever-larger fraction of the environmental challenges facing human-kind consists of problems requiring better management of human activity to reduce its deleterious impact on natural systems. Problems of this kind arise with increasing frequency because of the larger and more prosperous human population. But they can also be addressed with greater success because of our deeper understanding of the affected systems and an improved capacity to detect the impact of human beings. These kinds of problems come at all scales: from an individual room whose air is degraded by radon or organic pollutants, to an urban airshed subject to the buildup of pollutants intensified in particular seasons, to the global stratosphere, whose chemical composition is being altered by chlorofluorocarbons and nitrogen oxides.

The discovery of the destruction of stratospheric ozone by chlorofluorocarbons is a classic example of the use of physical science to understand how human beings change a natural system. Working out the details of this problem has involved a blend of the chemistry of heterogeneous reactions and the physics of fluids and radiation transport.

Global warming is partly a consequence of altering the carbon cycle on the planet by the burning of fossil fuels. The increase of carbon dioxide appears to foster the growth of other greenhouse gases by alteration of the global hydrological cycle. An understanding of global warming and the associated climate change draws on a number of disciplines. Geophysical fluid dynamics is necessary to understand the structure of the basic climate system within which these climate changes occur. At the same time, chemical and biochemical cycles are active partners in the dynamics and thermodynamics of the climate system.

Effective management of human interaction with an environmental system requires simultaneous progress on several fronts: an understanding of the system in the absence of human impact; an understanding of the way human impact changes the system; and an understanding of measures available to reduce this impact, such as substituting one form of energy production for another. Much progress has been made over the past few decades in understanding the workings of those environmental systems that are particularly vulnerable to human impact, ranging from the thermal behavior of lakes to the chemistry of the stratosphere. Many of these systems are now well understood, through a combination of measurement, modeling, simulation, and theory.

One of the best tools for measuring human impact on climate is the identification of small concentrations of tracer atoms in environmental samples. Various long-lived radioactive nuclei serve as such tracers in much the same way that short-lived radioactive nuclei serve as tracers for the study of biological systems. The use of these tracers has grown out of the understanding of the formation of radioactive elements and their decay and detection. This method for environmental monitoring has become increasingly important as ever more sensitive detection techniques are developed (see sidebar "Monitoring the Environment").

MONITORING THE ENVIRONMENT

Accelerator mass spectrometry (AMS) is an important tool for environmental measurements. AMS uses nuclear techniques to accelerate and identify small concentrations of tracer atoms in environmental samples. Measurements that would otherwise be difficult or impossible are made routine by its sensitivity.

Cosmic rays from elsewhere in the galaxy continually bombard Earth's atmosphere and surface, producing long-lived radioactive "cosmogenic nuclei." Because carbon in organic objects is not replenished from the atmosphere once an animal or plant dies, the ^{14}C present decays with a 5700-year half-life, and the amount remaining provides a measure of the object's age. Other cosmogenic nuclei can be used in a similar manner to determine how long material that contains them has been shielded from cosmic rays and from the atmosphere. The concentration of the long-lived isotope ^{81}Kr in an aquifer of the Great Artesian Basin in Australia is measured and used to determine how long its water has remained uncontaminated by younger groundwater.

Cosmogenic nuclei are used to study large-scale environmental phenomena. The amount of ^{10}Be in ice cores has been measured by AMS and is found to be correlated with solar activity. This correlation may allow studies of solar activity backward 10,000 years in time, compared to the 400-year record currently available. It may then be possible to determine to what degree solar variation is responsible for climate variation.

Other AMS measurements are devoted to understanding the nature of oceanic circulation, which has a major influence on climate. If northward-flowing currents in the Atlantic were to cease, the temperature in northern Europe would decrease by 5 °C to 10 °C. There is a concern that increasing greenhouse gases could initiate such a change. Measured concentrations of oxygen isotopes in Greenland ice cores show that large changes were common near the end of the last ice age. Dating of organic glacial remains in New Zealand using ^{14}C indicates that these large changes were global in nature.

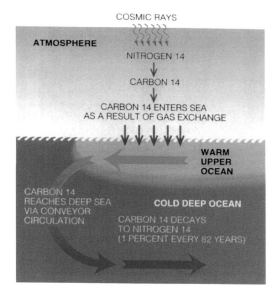

As cold, salty water sinks to great depths, as shown above, it carries radioactive ^{14}C out of the atmosphere and into the abyss, where it slowly decays. Radiocarbon dating is used to measure the state of the oceanic current system.

ENERGY PRODUCTION AND THE ENVIRONMENT

Improvements in energy efficiency contribute directly to environmental quality, and many of these improvements are applications of physics. Lighting efficiency has increased dramatically, progressing from the kerosene lamp to the incandescent bulb to the fluorescent light. Window thermal resistance, with its direct effect on the energy required for space heating and cooling, has been greatly increased by the use of thin-film coatings that embody insights from atomic physics. The oxygen sensor in the automobile exhaust, which permits far lower emissions of hydrocarbons, carbon monoxide, and nitrogen oxides, is another application of physics. Metal recycling is also in this category: The increasing competitiveness of secondary metals production, or recycling, relative to primary metals production resulted in part from numerous innovations in materials science.

The substitutability and attractiveness of alternative forms of energy production are also important. Fossil fuels inevitably produce carbon dioxide as a by-product of energy extraction. The two principal alternatives are nuclear power (both nuclear fission and nuclear fusion) and renewable energy (in many forms, including wind, hydropower, photovoltaic cells, and solar thermal energy). Research and development efforts on all of these alternative technologies are under way around the world. Each has its strengths and weaknesses. The sustainability of industrialized society is much more likely if humanity continues to pursue a broad portfolio of energy production options.

As long as fossil fuels continue to dominate energy production, the global energy system will require the sophisticated management of carbon. Carbon management strategies are already being considered and are attracting the attention of physicists around the world. An important class of strategies involves separating the carbon content from the energy content of fossil fuels by chemically processing the fossil fuels into carbon dioxide and hydrogen. The carbon dioxide would be treated as a waste product requiring sequestration from the atmosphere; for example, it might be piped to deep underground saline aquifers. The world would run its vehicles and buildings and appliances and factories on some combination of electricity and hydrogen, the two most environmentally preferred secondary energy carriers. The challenge of making a hydrogen economy safe and affordable will demand much progress in areas where physics is indispensable, ranging from molecular strategies for hydrogen storage to advanced materials for hydrogen fuel cells.

SUMMARY

Environmental science is highly interdisciplinary. The life sciences, chemistry, applied mathematics, geology, oceanography, and physics are all front and center. Physics plays a broad role, contributing directly to energy production and environmental projects and indirectly through basic research, providing technological spin-offs from research programs, and helping to educate a technically literate population capable of responding to environmental issues. Basic research in atmospheric and oceanic physics provides the foundation.

8

National Security

Physics and physicists are central to the nation's security. This partnership between government and physics includes important areas such as the design of optics for reconnaissance satellites, new forms of cryptography, the aging of the nuclear stockpile, communications electronics, counterterrorism, and ballistic missile defense. Rapidly evolving fields, such as the physics of new materials, and various applications of physics, ranging from physical oceanography to remote sensing, now are crucial for national security.

Many of the technical breakthroughs that have contributed to national security have their roots in advances in basic physics research. Recent military actions in the Gulf War and in the Balkans showed the extent to which warfare has been transformed by technology. Technical superiority shortened the duration of these actions and helped to minimize the loss of life. Physics was involved directly, as in laser guidance and satellite technology, and indirectly, by virtue of the many areas of basic research that underpin modern electronics, optics, and sensing systems. Scientists engaged in basic research also play a crucial role in evaluating new threats and opportunities arising from technical advances. Scientific risk/opportunity assessment increases the chances the nation will invest its defense resources wisely and avoid reactions to misperceived threats.

Physics research relevant to national defense involves a host of agencies, university connections, and industrial links; each could be the subject of a dedicated report. The committee focuses here on two broad areas that touch on many of these fields: basic physics research at the national laboratories of the Department of Energy and at the Department of Defense.

THE DEPARTMENT OF ENERGY

An important national legacy of the government-science partnership that grew out of World War II is the national laboratories operated by the

Department of Energy's Office of the Deputy Administrator for Defense Programs (hereinafter called the Office of Defense Programs): Los Alamos, Livermore, and Sandia. These laboratories today have a central mission—reducing the global nuclear danger—that involves extraordinary challenges in stockpile stewardship, in nonproliferation and arms control, in nuclear materials management, and in the cleanup of the environmental legacy of nuclear weapons activities. They have also shouldered other important responsibilities as the government has recognized new issues that affect the nation's physical and economic security and that require technological solutions. Examples include global climate dynamics, new energy sources, counterterrorism (including chemical and biological weapons of mass destruction), environmental protection and remediation, and biomedical technologies.

To succeed in their missions, it is essential that the laboratories have access to excellent science and technology. The core technical competencies that have been established in crucial areas such as high-energy-density physics, nuclear physics, hydrodynamics, computational science, and advanced materials are the cornerstones supporting the laboratories. The facilities and scientific manpower concentrated in these areas are the result of years of government investment. The laboratories are also supported by a network of contacts to the outside world, including university researchers, industrial partners, and Department of Defense scientists. These interactions are important both in leveraging scientific strength and in recruiting new talent to the laboratories.

The Laboratories and Global Nuclear Dangers

The invention of nuclear weapons was one of the defining events of the 20th century. The political and military legacy of this invention is now exceedingly complex. The underlying physics of atomic weapons is widely understood and accessible to the scientists of many nations. Indeed, the original technology dates to more than 50 years ago, when most movies were black and white, telephones needed operators, and radio had not yet been supplanted by television. Fissile material, once a great barrier to entering the nuclear community, now exists in great quantities in the United States, the former Soviet Union, and elsewhere. Estimates of this stockpile range from 100,000 to 1,000,000 kg, while the amount needed for a bomb is about 10 kg. There is great concern that not all of this fissile material is confined to politically stable parts of the world.

The rapid lowering of the barrier to the nuclear club led the United

States and other nations to enter into the Comprehensive Nuclear Test Ban Treaty to limit the spread of nuclear weapons. The effort to stem proliferation while maintaining national security poses new challenges to the Office of Defense Programs' national laboratories, one of them being that the most advanced weapons in the U.S. arsenal are now about a decade old. The laboratories have the congressionally mandated duty of verifying the readiness and reliability of the weapons stockpile, as well as responsibility for maintaining the capability to resume underground testing, to execute new designs, and to understand the nuclear weapons capabilities of other nations.

In the absence of nuclear testing, the laboratories rely increasingly on laboratory experiments and computer simulations to predict the behavior of weapons. Uncertainties that previously could be handled empirically—through ad hoc adjustments of codes to reproduce the results of tests—must now be handled quantitatively, through ancillary physics experiments or improved theory. The physical data effort is being aided by high-energy-density research devices such as the Z-pinch at Sandia and the Omega laser at the University of Rochester, in which conditions almost as extreme as those produced in the explosion of a nuclear weapon (or a supernova) prevail. Another step will have been taken with the completion of Livermore's National Ignition Facility, which will exploit lasers to study the high energy densities required for laboratory nuclear ignition experiments.

Appropriate modeling of the physics and excellent numerical resolution in time and space are both essential to realistic simulations. The Accelerated Strategic Computing Initiative (ASCI) is an effort to exploit fully the power of massively parallel computing to model and verify weapons performance. It challenges computer scientists to utilize effectively new parallel architectures and physical scientists to model properly the complex physical processes that govern weapons behavior. This effort will have impacts well beyond weapons: Other problems of national interest—among them the efficiency of internal combustion engines, climate and weather modeling, and the spread of forest fires—involve similar numerical and physics challenges.

The effects of aging on the stockpile present another class of challenges. There are important new techniques on the horizon—for example, proton radiography—that promise to help scientists monitor nondestructively the changing properties of weapons materials. The ability of the laboratories to develop this technology is a direct consequence of their long involvement in accelerator physics and detector technologies.

Security and Basic Research

Security is of paramount importance to the defense activities of the Office of Defense Programs' national laboratories. This need is an additional challenge for laboratory scientists, as science is a collective endeavor in which discussion and open criticism speed progress and are a central part of the process of validation. Such open exchanges can conflict with the need to compartmentalize knowledge for security purposes. The scientists at the laboratories thus often have to pursue their science under conditions that restrict the feedback they receive. The laboratories have recognized this issue and work hard to provide the needed peer review under conditions consistent with security.

Laboratory security is sometimes a contentious issue. There are concerns on the one hand about the adequacy of security practices and on the other about reactions to security breaches that will isolate the laboratories from the outside scientific community. Increased isolation will diminish the quality of science at the laboratories and detract from the recruiting and retention efforts needed to keep these institutions strong for another generation.

Richard Rhodes, in his account of the development of the atomic bomb, attributed the success of the U.S. effort—as well as the slower progress of the Soviet effort during World War II—in part to the degree of trust the respective governments had in the judgment of their scientists.[1] This constructive relationship between weapons laboratory scientists and government has persisted and served the nation well for nearly 60 years. While security is essential to the nation's defense programs, it is also important for Congress, the Department of Energy and its laboratory leaders, and laboratory scientists to work together to ensure an atmosphere of trust.

Preserving Laboratory Quality

The technical strength of the Office of Defense Programs' laboratories derives from the quality of their scientists and of their facilities. There are troubling trends that threaten to weaken both.

The basic science activities within the laboratories—those activities that maintain the core competencies and provide much of the innovation—appear to be in significant decline at Livermore, Los Alamos, and Sandia. This decline has been driven by rather dramatic changes in the way the

---

[1]Richard Rhodes. 1995. *The Making of the Atomic Bomb.*

laboratories are funded: Increasingly, support is directed narrowly to specific programmatic efforts. This is a departure from past practices, in which a portion of short-term programmatic funding was reserved for the support of core science efforts important to the long-term health of the laboratory. At Livermore, for example, funding of the Physics Directorate has declined by 30 percent in 3 years. This has led to the closing of a number of smaller facilities that previously helped to provide the physical data needed for weapons design. It is a troubling trend given that basic science and physical data should be of increasing importance to stockpile stewardship because they are necessary input for the simulation efforts.

The impact of these reductions was heightened in FY00 by a reduction in laboratory funding for start-up basic research by one-third. Fortunately, Congress restored this funding to its usual level, 6 percent of laboratory budgets, in FY01. These funds have allowed laboratory scientists to pursue new basic research directions and to identify new programmatic possibilities. Laboratory leaders have recognized the importance of keeping some of their most creative scientists thinking about the nation's future needs even as the pace of technology development requires heightened vigilance.

Another threat to national security is the growing difficulty facing laboratory recruiters. Two important national trends, the decline in the number of U.S. physical science Ph.D.'s and the increasing competition from industry for the best young scientists, would present a problem for the laboratories even in the best of times. The effects of these trends are now compounded by morale and funding issues. There has been a significant drop-off in applications to the Office of Defense Programs' national laboratories. The impact on applications from non-U.S. citizens, a major fraction of the talent pool, appears to be especially severe, a decline by a factor of about 5. Low morale is affecting long-term employees as well, making outside opportunities appear more attractive. As the best scientists in the national laboratories are clearly also the most marketable and most mobile, there is a risk that talent will rapidly be lost.

Many of the scientists with experience in weapons design and testing have retired or will retire soon, making their replacement an immediate issue. The laboratories must recruit young scientists from the pool of researchers produced by our leading universities. Historically, recruitment has been greatly enhanced by the strong basic science efforts of the national laboratories in core competencies like astrophysics, nuclear physics, high-energy-density physics, hydrodynamics, computer science, and atomic physics. These areas draw large numbers of new scientists to the laboratory, many of whom later become fully involved in activities of a more programmatic kind. As the weapons program basic science support and laboratory basic science start-up funds dry up, so too does the conduit for drawing new talent into the laboratories.

THE DEPARTMENT OF DEFENSE

In the decades following World War II, the Department of Defense supported a broad portfolio of basic research at DOD, university, and industrial laboratories. The motivation was the expectation that technical advances would further the nation's capabilities in areas such as surveillance, intelligence gathering, missile defense, communications, stealth technology, and nuclear physics. The Navy has interests in oceanographic physics, in the propagation of sound through water, in deep-ocean currents, and in meteorology. Air Force concerns include turbulent fluid flows, navigation, long-range observation, and pattern recognition, while Army interests include night and all-weather vision and techniques for avoiding detection. The Air Force, Navy, and Army share many common goals: Each service depends on surveillance and reconnaissance to assess threats before battle and to follow the evolution of a conflict once battle is joined; all need to defend their positions and to locate targets and destroy them before they themselves are attacked.

But defense research is changing. With the end of the Cold War and the diminished threat of all-out conflict between the United States and Russia, DOD budgets have fallen and a very different world of threats faces the military—smaller conflicts, multiple conflicts, biological and chemical threats, terrorism, cyberwar. Many other changes accompany these threats. The new information-based economy has altered the relationship between DOD development and industry. More off-the-shelf products find their way into weapons systems, driven either by cost considerations or by the rapid advance of commercial capabilities, which often outstrip what government can do on its own. Finally, the move toward multidisciplinary, integrated systems of data and control has joined physics to other arenas in fundamentally new ways, with nanotechnologies as one instance and the computational sciences as another.

Physics and the DOD

The modern battlefield has changed remarkably as a result of technological advances. Lasers guide smart munitions and help in high-resolution surveillance. Advanced optical systems are employed in space-based satellite surveillance systems, in manned and unmanned aircraft, in missiles, and even on rifles. In the Gulf War, night vision systems proved to be a crucial technology (Figure 8.1). Forward-looking infrared detectors (FLIRs) are now being acquired in the hundreds of thousands. There have been rapid developments in areas such as directed-energy weapons, surveillance, stealth, electronic countermeasures, guidance and control, information and signal processing, communications, and command and control. The pace at which a weapons system proceeds from the conceptual, to the commonplace, to the obsolete continues to accelerate.

FIGURE 8.1   A soldier at Fort Benning, Georgia, tests the latest Land Warrior gear. Land Warrior, the Army's newest weapons system, provides improved situational awareness, high levels of protection, rapid digital and voice communications, and accurate targeting using thermal and video sighting. Image courtesy of U.S. Army Soldier Systems Center.

These technologies depend on underlying advances in a wide range of physics disciplines. Many of the examples above reflect recent developments in optics. Plasma physics figures widely, from beam weapons to display technologies. Quantum physics is the foundation for novel electronic devices and components. Atomic and molecular physics figures in clocks for navigation, the Global Positioning System (GPS), lasers, and the observation of atomic interactions in strong electromagnetic fields. These connections to physics motivated the strong DOD investments in basic science during the Cold War. Physicists in DOD laboratories not only contributed to the advance of basic science but also helped DOD to keep abreast of and evaluate the relevance of developments in industry and in universities.

DOD Support of Basic Research

The DOD divides its research cycle into a series of budget categories from 6.1 to 6.7, categories that, however imperfectly, are designed to track research funding from the most fundamental to the processing of operational weapons systems. Funding for basic research (6.1) along with other defense spending began a decline at the end of the Cold War. Measured in constant FY01 dollars, this decline took the executed 6.1 budget from roughly $1.49 billion in 1993 to a low of just over $1.06 billion in 1998; since then, there has been a rise to approximately $1.17 billion in FY00 and to $1.33 billion in FY01. This represents a decline of approximately 11 percent over the period. DOD support of basic research in physics has moved in step with the overall research budget since the end of the Cold War, also decreasing by approximately 11 percent, bringing it to just over $122 million in FY01.

DOD basic research now represents approximately 6.5 percent of the total federal commitment; by comparison, the DOE budget for basic research, at $2.3 billion, represents about 13 percent. The DOD support is provided through the Army, the Navy, the Air Force, and the Office of the Secretary of Defense, with the remainder distributed among other defense agencies and the Defense Advanced Research Projects Agency (DARPA). At least 50 percent of the DOD basic research budget goes to universities, about 25 percent to in-house DOD laboratories, and the remaining 25 percent to an assortment of industrial and other sites. In certain sectors of research, DOD funding represents a powerful component of the total federal support at universities. DOD funding, for example, now accounts for about 70 percent of all the federal funding for electrical engineering. Computer science gets nearly 50 percent and mathematics gets 17 percent of its federal funding through the DOD.

SUMMARY

Changes in recent years occurring in the national laboratories operated by the Department of Energy's Office of Defense Programs may be altering a formula that has served the nation well for half a century: National security challenges are best addressed by laboratories with excellent basic science core competencies and with strong connections to outside university and industrial researchers. The dominant force behind the changes is a pattern of funding that de-emphasizes the long-term basic research that previously maintained laboratory excellence in core competencies. In addition, unfortunate security lapses and the response to them are contributing to morale and recruiting problems, endangering the historical partnership between government and laboratory scientists.

Some scientific leaders in DOD feel that budget reductions for basic science have seriously weakened in-house research: Long-term declines and year-to-year instabilities have made it difficult to retain the top scientists. Instabilities in DOD external funding of industry and university research have also resulted in considerable disruption of programs with a corresponding loss in productivity.

The decline of DOD laboratory basic science capabilities and activities raises issues similar to those raised at the DOE laboratories. In earlier decades, the DOD laboratories had active programs in basic physics research directly relevant to DOD missions. The scientists involved in such research were able to advise the DOD on basic physics issues and to help evaluate products provided by industry. The decline in this research effort and in the quality of in-house expertise has been driven by changing funding trends, short-term demands on the services' budgets, and competition for good technical people from other sectors of the economy.

There is a critical need to ensure that the physics research required to maintain the technical superiority of the nation's armed forces is being carried out somewhere. Although the committee is not in a position to judge whether or not the DOD laboratories are the best places to do this, it is clear that they had this function in the past and now have lost much of their capability. Regardless of the source of DOD research, there is also a critical need for the DOD to evaluate the physics carried out by outside vendors. The level of DOD in-house expertise may no longer be sufficient for this task.

9

The Economy and
the Information Age

We live in an era of astounding technological transformation in which change, not stability, has become the norm. All around us are now-familiar technologies whose present state of development—or very existence— would have seemed extraordinary just a generation ago. From wireless telephones and handheld GPS units to video games, DVD, and digital television; to genetic engineering, decoding of the human genome, and combinatorial drug design; to MRIs, CT scans, laser eye surgery, and robotic hip replacement surgery; to polymeric materials for inline skates, tennis rackets, and skis; to superalloys for jet engine turbine blades; to lightweight, fuel-efficient automobiles and aircraft; to computers, the Internet, and the World Wide Web—technology is everywhere and touching all of us in ever more pervasive ways (see sidebar "The World Wide Web").

All of this technology does not just happen. It arises from innovation based on decades of research in the underlying basic science, of which physics is a major component in essentially every case. Medicine, entertainment, transportation, energy, national security, communications, computers—the unprecedented advances in all of these areas are built on solid technical foundations resulting from a half-century and more of continuous investment in basic science. It is the investment in basic science that we are making today that will fuel the technological innovations of tomorrow (see Figure 9.1).

At the heart of this age of change is information technology. We are in the midst of an information revolution that is every bit as profound as the two great technological revolutions of the past—the agricultural and industrial revolutions. Information technology sector revenues are estimated to account for 5 to 15 percent of the U.S. GDP, and 40 percent of U.S. industry capital spending today is for information technology.

Scientific understanding of fundamental phenomena has been key to the development of materials for the information age, the carriers and con-

THE WORLD WIDE WEB

By now most Americans are well aware of the World Wide Web, and for many it has become an integral part of their business and recreational lives. It is creating new business processes and models, from book sales to airline tickets to banking to vacation planning to stock trades to home and automobile purchases. Few of its users, however, are aware of how it came to be.

The Web was born in a high-energy physics laboratory in Switzerland. It came about as a solution to CERN's communications and documentation problems, characteristic of large, complex experiments involving collaborators from around the world. Three key technologies—computer networking, document/information management, and software user interface design—were brought to bear on these problems, and the outcome became the Web.

The Internet, initiated by DARPA in the late 1960s, was well entrenched at CERN by the late 1980s. At that time it was expanding explosively worldwide, due in large part to the wide acceptance of e-mail as an effective means of communication. The Internet became the medium in which the Web was created.

In the fall of 1990 the proposal for the World Wide Web, including its name, was advanced and acted favorably upon at CERN. Its eight key goals, defining features of the Web, were as follows:

• To provide a common (simple) protocol for requesting human-readable information stored at a remote system using networks;

• To provide a protocol within which information can automatically be exchanged in a format common to the supplier and the consumer;

• To provide some method of reading at least text (if not graphics) using a large portion of the computer screens in use at CERN;

• To provide and maintain at least one collection of documents, into which users may (but are not bound to) put their documents. This collection will include much existing data;

• To provide a keyword search option, in addition to navigation by following references, using any new or existing indexes. The result of a keyword search is simply a hypertext document consisting of references to nodes that match the keywords;

• To allow private, individually managed collections of documents to be linked to documents in other collections;

• To use public domain software wherever possible, or interface to proprietary systems that already exist; and

• To provide software for the above free of charge to anyone.

In a little over a year, a complete version of this proposal had become a reality and announced to the high-energy physics community. In the process, HTTP (Hypertext Transfer Protocol), which allows the client and server to communicate, was invented, along with HTML (Hypertext Markup Language, based on SGML), which allows content to be displayed on a client.

The initial acceptance of the Web by the high-energy physics community was far from unanimous. It was finally the browser interface to the SPIRES (Stanford Public Information Retrieval System) databases at the Stanford Linear Accelerator Center, containing a wide range of information on high-energy physics experiments, institutes, publications, and particle data, that sold the Web to that community. Subsequently, the Mosaic browser developed in early 1993 at the National Center for Supercomputing Applications at the University of Illinois set the Web on the path to the broad role it plays today. By the end of 1995 there were an estimated 50,000 Web servers worldwide, up by 20-fold over a 1-year period.

Over the past 5 years the World Wide Web has continued to grow exponentially. Today there are hundreds of millions of Web servers worldwide, with e-commerce and the proliferation of dot-coms emerging throughout the business world. The Web has had a truly astonishing trajectory over the 10 short years from its inception in a physics laboratory to the front burner of our nation's and the world's economy.

Tim Berners-Lee, founder of the World Wide Web, at CERN.

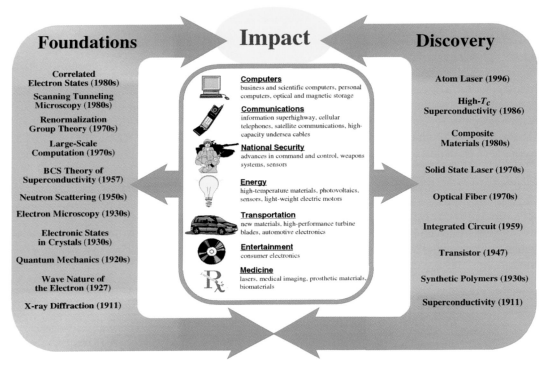

FIGURE 9.1   The incorporation of scientific advances into new products can take decades and often follows unpredictable paths. The physics discoveries shown in this figure have enabled breakthrough technologies in virtually every sector of the national economy. The most recent fundamental advances leading to new foundations and discoveries have yet to realize their potential.

trollers of electrical current, light waves, and magnetic fields. Equally important is a scientific understanding of the processes, because it is such understanding that enables the lower-cost manufacture of devices and systems based on materials used by the rapidly changing electronics and telecommunications industries. This last point cannot be overemphasized. Without the continual decrease in the cost of processing, communicating, and storing bits of information, the information age would never have happened, irrespective of the sophistication and elegance of the technology. Figure 9.2 shows how much computational capability $1000 was able to buy over the past half century, demonstrating faster-than-exponential growth of approximately nine orders of magnitude during that time. Through basic research in physics, the technology itself has changed completely several

times during this same 50 years, with the integrated circuit having domi-
nated since the mid-1970s.

INTEGRATED CIRCUITS

Semiconductors have grown into a global industry with year 2000 revenues of about $200 billion, supported by a materials and equipment infra-
structure of about $60 billion. Semiconductor technology is also the heart of
the $1 trillion global electronics industry and is vital in many other areas of
the approximately $33 trillion global economy.

The predominant semiconductor technology today is the silicon-based
integrated circuit. The key foundations of the modern silicon semiconductor
industry are the discovery of the electron in 1897, the concept of the field-
effect transistor (FET) in 1926, the first demonstration of the metal oxide

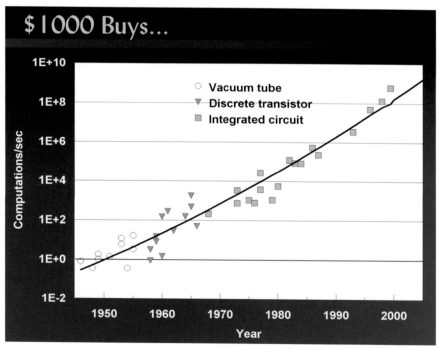

FIGURE 9.2   Computational power that $1000 has bought over the last 50 years.
SOURCE: Hans Moravec. 1998. "When Will Computer Hardware Match the Human
Brain?" *Journal of Transhumanism*, Vol. 1.

semiconductor FET in 1959, and the development of dynamic random access memory (DRAM) in 1967 and the first microprocessor in 1971. In 2000, the Nobel Prize in physics was awarded in part for basic and applied research leading to the integrated circuit.

For the past 30 years, semiconductor technology has been described by Moore's law, an empirical observation that the density of transistors on a silicon integrated circuit doubles about every 18 months. Today's computing and communications capability would not have been possible without the phenomenal exponential growth in capability per unit cost since the introduction of the integrated circuit in about 1960. That sustained rate of progress has resulted in high-density DRAMs with 64 million bits on a chip and complex, high-performance logic chips with more than 9 million transistors on a chip. Figure 9.3 illustrates the exponential growth in micropro-

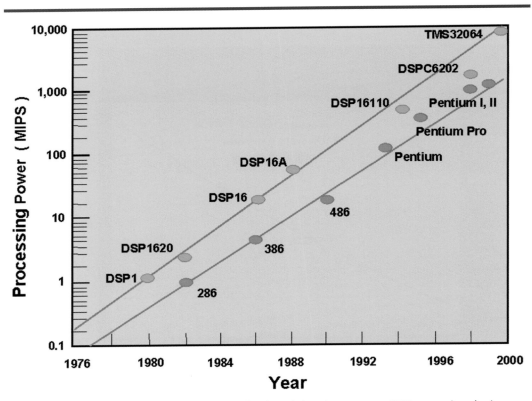

FIGURE 9.3 Computational power versus time for digital signal processors (DSPs, green) and microprocessors (red). Courtesy of Intel and Texas Instruments.

cessor computing power over the past 25 years. The availability of enormous computational power at low cost is having a dramatic impact on our economy. In addition, it feeds back directly into the scientific enterprise and is stimulating a third branch of research in physics, computational physics, which is rapidly taking its place alongside experiment and theory.

Moore's law, first articulated by Gordon Moore of Intel Corporation, is not a physical principle. It is, instead, a statement that industry will perform the R&D necessary and supply the required capital investment at the rate required to achieve this exponential growth rate. While this has certainly been the case so far, there is considerable debate over how the future will play out. Exponential growth cannot continue forever, and at some point, for either technical or economic reasons, the growth will slow. Many daunting scientific and engineering problems must be overcome for industry to continue at the Moore's law rate of progress for the next 15 or even 10 years. For instance, the number of wires needed to connect the transistors grows as a power of the number of transistors. As transistor dimensions are shrunk, integrated circuit manufacturers pack an ever-increasing number of devices into their chips. The complexity of wiring the transistors in these chips may eventually reach the limits of known materials. The cost of manufacturing increasingly layered and complex wiring structures may limit the performance of these systems. Even if solutions to the interconnect problem can be identified, continued scaling of silicon technology will ultimately encounter fundamental limits. For example, metal-oxide semiconductor transistors can be built today with gate lengths of 30 nm (about 150 atoms long) that display high-quality device characteristics. Manufacturing complex circuits that rely on devices with features of this size will require several hundred processing steps with atomic-level control. Moreover, the performance of complex integrated circuits with tens of millions of transistors may be degraded because of nonuniform operating characteristics. In time, continued decreases in device dimensions may result in the information being carried by an ever-decreasing number of charge carriers; ultimately, simple statistical fluctuations will limit the uniformity of device characteristics as the number of charges used to convey information decreases.

To delay the onset of these limits as long as possible, research is under way on new materials with a high dielectric constant appropriate for both memory applications and for limiting the leakage of current in tightly packed integrated circuits. As the understanding of synthesis and processing increases, ferroelectric materials are being introduced for nonvolatile memory applications. Even with these advances, as feature sizes continue to decrease, integrated circuits based on field-effect transistors will eventually encounter fundamental limits such as interconnect delays caused by the ever-increasing number of interconnects, heat generation, or the quantum limits of transistors too small to confine the electrons in the conduction channels. Ultimately, devices that rely on the manipulation of single electrons may play an important role. Today's approach to the design and manufacture of integrated circuits, which already requires processing and control at near single-atom levels, will no longer be extensible to smaller feature sizes and higher densities.

OPTICAL-FIBER COMMUNICATION

If silicon integrated circuits are the engine that powers the computing and communications revolution, optical fibers are the highways for the information age. Roads and highways changed in the last century to accommodate the explosion of cars and trucks. Multiply that by a million and you'll get an idea of the scope and the pace of change of the communications revolution. Today's communications networking industry is one of the most dynamic and rapidly changing industries in the world. It is expected that nearly $1 trillion will be spent in the next 3 years on building the next generation of Internet networks. It took nearly a century to install the world's first 700 million phone lines; 700 million more will be installed in the next 15 years. There are more than 200 million wireless subscribers in the world today; another 700 million will be added in the next 15 years. There are more than 200 million cable TV subscribers in the world today; 300 million more will be added in the next 15 years. All of this would be impossible using the copper wires that were the mainstay of the telecommunications industry only 20 years ago.

Optical-fiber communication is based on a number of developments in basic physics and materials science. They include the purification and processing of tiny pipes of glass, called optical fibers, that efficiently guide light over long distances with so little absorption or scattering that the light can travel for hundreds of kilometers without being absorbed; the invention of the laser and its realization in tiny, efficient semiconductor chips, which gives us a suitable source of the light; and, finally, the invention of the erbium-doped fiber amplifier, which allows efficient amplification of the optical signals.

An erbium fiber amplifier is an optical fiber that has a small amount of exotic material (the element erbium) mixed in with the glass of the fiber.

When irradiated by light from very recently invented solid-state lasers, erbium atoms act as amplifiers for the light signals through the fiber, boosting the strength all at once to compensate for any losses. This replaces complex, expensive, and relatively unreliable repeater stations, which must separately amplify each channel by absorbing the light from a single channel on a detector, processing the signal, then feeding it to another laser that emits light to send the signal on its way. Since such stations are difficult and expensive to replace—they are often at the bottom of the ocean—the advantages of a simple, highly reliable system of amplification are obvious. These amplifiers would never have been possible without basic research to understand what colors of light are absorbed and emitted by different atoms, how the absorption and emission of light for specific atoms are affected by the insertion of the atoms into the glass host, and how the atoms can be excited by one color of light to act as an amplifier for another.

Fiber amplifiers were first used because of their simplicity and reliability. They have also resulted in a major improvement in the amount of information that can be transmitted down a single fiber. This is shown by the sudden change in the slope of the curve for the information capacity of a single fiber (in 1993 and 1997 for experimental and commercial fibers, respectively), as shown in Figure 9.4. Because fiber amplifiers amplify the light without changing it, they amplify different colors of light at the same time. Multiple colors of light can now be sent simultaneously down the same fiber, each carrying its own information, much like the multiple channels of radio and TV signals. This so-called wavelength division multiplexing (WDM) is being used to multiply the effectiveness of currently installed fiber many times over. A few years ago, optical physicists discovered that it was possible to change slightly the properties of glass under particular conditions by irradiating it with ultraviolet light. This has made possible the fiber gratings that are now used to combine and separate the different colors of light that make up the different WDM channels. Also, basic research on atoms and molecules that involved precisely measuring the color of light that they absorb is now used to stabilize the color of light used for each WDM channel, allowing even closer spacing of channels that do not interact, called dense WDM (DWDM).

In part because of the faster than exponential growth of connections to the Internet, optical fiber is being installed worldwide at the rate of more than 20 million km per year—more than 2000 km every hour. Optical telecommunication was introduced into the market in 1980; today, not only is optical fiber the medium of choice for long-distance voice and data

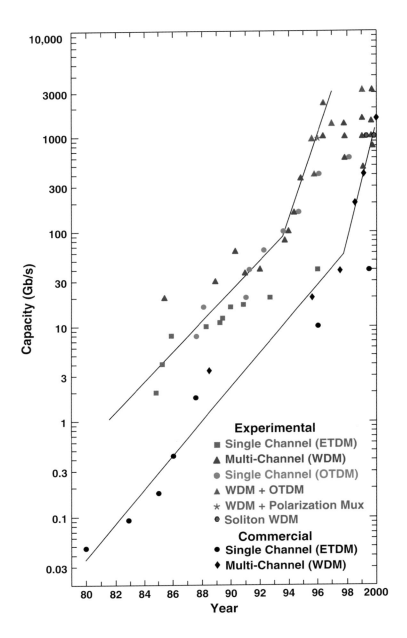

FIGURE 9.4   Optical-fiber capacity, 1980 to 2000. The development time between the demonstration of a specific experimental system and its commercial deployment has narrowed from about 5 years to just a few months over the last two decades. Courtesy of Bell Laboratories, Lucent Technologies.

communications, but it is also rapidly growing to be a leading player in the local area network (LAN) market. Optical-fiber manufacturing revenues are predicted to be about $30 billion in 2003.

The rate of information transmission down a single fiber is increasing exponentially. Transmission at 3 terabits per second (Tbps) has been demonstrated in the research laboratory, and the time lag between laboratory demonstration and commercial system deployment has shortened to several months, as shown in Figure 9.4. What cannot be as easily discerned from the figure is the fact that the analogue of Moore's law for fiber transmission capacity, which serves as a technology roadmap for lightwave systems, is a doubling every 9 months (twice as fast as the rate for transistor density on a silicon chip).

The first undersea optical cable, installed in 1988, had a capacity of about 8000 voice channels per cable at a cost of about $400 per channel. More than 300,000 km of undersea lightwave cable had been installed by the end of 1996, when it cost less than $30 per year per voice channel and had 120,000 voice channels per cable (5 Gbps per fiber). The first large terrestrial lightwave system installed in the United States linked Washington, D.C., and New York City with a capacity of 90 Mbps per fiber in 1983, and a similar system linked New York and Boston in 1984. More than 230 million km of fiber had been installed worldwide, about two-thirds of it in the United States, by the third quarter of 2000. The latest systems incorporate WDM, dispersion-shifted fiber, and optical amplifiers. Currently in deployment are 400-Gbps-per-fiber systems using 40 channels with 10 Gbps per channel. In the next year or two, 3- to 6-Tbps-per-fiber systems will be introduced into the market, with 40 Gbps per channel. The ultimate information theory limit to the amount of data that can be transmitted over a single fiber is estimated to be on the order of 20 Tbps and is currently a subject of intense study by physicists.

As with silicon technology, it is also the dramatic reduction in cost that has driven fiber technology development so rapidly. Figure 9.5 shows the tenfold decrease over the last 5 years in the cost of transmitting data over 1 km of fiber. This cost reduction has continued for two decades. This sustained reduction in the cost of communication has been a key driver of the information revolution in general and the Web in particular (see also the sidebar "MEMS for Optical Switching and High-Density Storage"). It is specifically the reduction in the cost of communication that is fueling the growth of the World Wide Web.

The photons that transport information along optical information highways are provided by compound semiconductor diode lasers. Such lasers

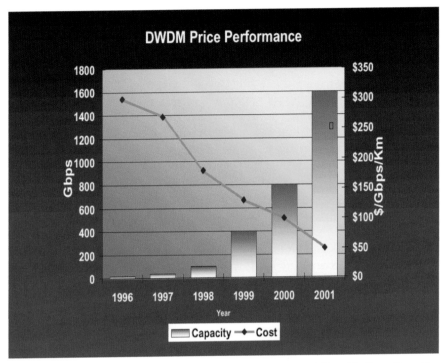

FIGURE 9.5   Total capacity per fiber is increasing exponentially with time. Over the last 5 years the cost of transmitting a gigabit-per-second signal for 1 km of fiber has decreased by a factor of 10. Courtesy of Bell Laboratories, Lucent Technologies.

are also at the heart of optical storage and compact disc technology. Because compound semiconductors have two or more different atomic constituents, they can be tailored by selecting materials that have the desired optical and electronic properties. Exploiting decades of basic research in materials such as gallium arsenide and indium phosphide, we are now beginning to be able to understand and control all aspects of compound semiconductor structures, from mechanical through electronic to optical, and to grow devices and structures with atomic layer control. This capability allows the manufacture of high-performance, high-reliability lasers to send information over the fiber-optic networks. High-speed, compound-semiconductor-based detectors receive and decode this information. These same materials provide the billions of light-emitting diodes sold annually for lighting displays, free-space or short-range, high-speed communication, and

MEMS FOR OPTICAL SWITCHING AND HIGH-DENSITY STORAGE

With the advent of microelectromechanical systems (MEMS) micromirror technology for routing light signals, we are entering into the era of all-optical networks. MEMS micromirror arrays such as the one depicted below at the left enable an optical cross-connect configuration setup to switch any light signal in any format at any bit rate coming into any of a thousand input fibers into any of a thousand different output fibers without the costly conversion of optical to electrical and back to optical signals that exists in today's networks. Each tiny gold-coated silicon mirror in the figure is about a millimeter in diameter. Mirrors can rotate along two axes on tiny gimbal mounts, bouncing a light beam that hits the mirror face in any direction. Optical switches, along with optical amplifiers, all-optical repeaters, and wavelength converters, will create the all-optical networks of the future, which will be both more flexible and less expensive for the consumer. These technologies are all expected to be emerging in the near future from the research laboratory after decades of fundamental research in materials science.

The areal density limit for magnetic storage, as currently practiced, is believed to be about 100 Gbit/in.^{2}, set by thermal stability limitations in the media. To continue the rapid growth in areal density beyond this limit, a number of different approaches are being considered. One approach is based on atomic force microscopy (AFM). The basic idea is to store information in the form of physical structure on a surface, which is read out as ones and zeros with an AFM. In the example below at the right, the structure takes the form of dimples that are thermomechanically written in a film with the same AFM that does the reading. A drawback of this approach is that the read/write bandwidth of a single AFM is low owing to its relatively slow scan rate. To overcome this limitation, an array of 103 or even 106 AFMs, all integrated on a single MEMS chip, is being investigated. The chip is scanned as an array of tiny phonograph needles over the media, with all the AFMs operating in parallel to read out the data. This approach holds promise of achieving storage densities of several terabits per square inch or more with wide bandwidth read/write capability.

Lightwave Micromachine Optical Crossconnect

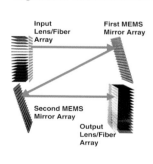

Input Lens/Fiber Array

First MEMS Mirror Array

Second MEMS Mirror Array

Output Lens/Fiber Array

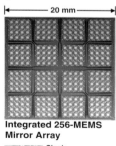

|← 20 mm →|

Integrated 256-MEMS Mirror Array

Single Mirror

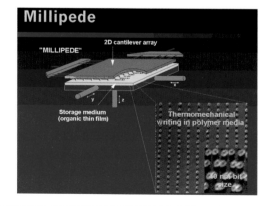

Millipede

"MILLIPEDE"

2D cantilever array

Storage medium (organic thin film)

Thermomechanical writing in polymer media

numerous other applications. In addition, very-high-speed, low-power compound semiconductor electronics in materials such as gallium arsenide and—recently—silicon, germanium, and gallium nitride plays a major role in wireless communication, especially for portable units and satellite systems. The compound semiconductor industry has approximately $1 billion in revenues today and is growing at 40 percent per year.

INFORMATION STORAGE

The third key enabler of the information revolution is low-cost, low-power, high-density information storage that keeps pace with the exponential growth of computing and communication capability. While both magnetic and optical storage are in wide use, computer hard disk drives, which use magnetic storage, constitute the single largest market, about $30 billion per year.

The highest-performance magnetic storage/readout devices have now begun to rely on giant magnetoresistance (GMR), the highly sensitive dependence of the electrical resistivity of certain metallic multilayered structures on magnetic field. Although Lord Kelvin discovered magnetoresistance in 1856, it was not until the early 1990s that commercial products using the technology were introduced. In the past decade, our growing understanding of condensed matter and materials led to important advances in our ability to deposit materials with atomic-level control, enabling production of the GMR heads that were introduced in workstations in late 1997. This increased understanding, made possible by our growing computational ability coupled with atomic-level control of materials, has led to exponential growth in the storage density of magnetic materials. From the mid-1960s through about 1990, the compound growth rate in areal storage density was about 25 percent per year. In about 1990, the rate increased to 60 percent per year, coincident with the introduction of magnetoresistive heads. As shown in Figure 9.6, there has recently been another increase in the rate to 100 percent per year, coincident with the introduction of GMR heads. Also shown in the figure is the reduction in time between initial demonstration and first shipment of the product. This is an unambiguous indication of the urgency and intensity with which information storage is progressing.

Like the silicon and fiber-optic industries, the storage industry has been driven by remarkable reductions in cost. Figure 9.7 shows the cost of 1 MB of hard disk drive storage capacity over the past 20 years. This cost has gone down by three to four orders of magnitude since 1980. For comparison, the cost of semiconductor memory is also shown.

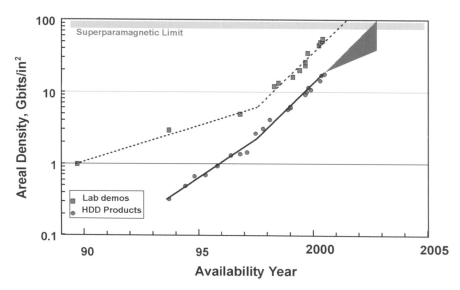

FIGURE 9.6   Hard disk drive (HDD) areal density is now doubling every 12 months. Courtesy of IBM Research.

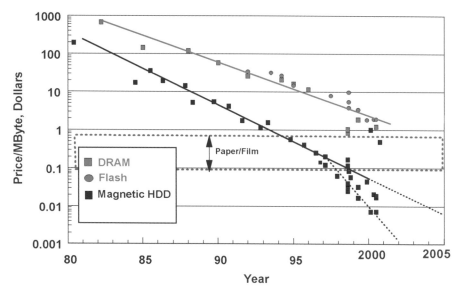

FIGURE 9.7   Cost per megabyte of magnetic hard disk drive (HDD) storage, 1980 to 2005 (projected). Semiconductor memory (Flash and DRAM) is shown for comparison. Courtesy of IBM Research.

It is amazing that while magnetic effects drive a $100 billion per year industry, our basic understanding of magnetism, even in a material such as iron, is incomplete. The fundamental limit on the stability of magnetic domains is an important area of basic investigation in magnetism. Materials exhibiting colossal magnetoresistance (CMR), with much more sensitivity to magnetic fields than even GMR, are being actively investigated. The advancing march of magnetic technology makes investigation of these limits inevitable, but these are also some of the most challenging questions for condensed-matter physics and materials science. What is the smallest-size magnetic element that is stable against external perturbations such as temperature fluctuations? Given that quantum mechanics sets bounds on the lifetime of any magnetic state, how do such bounds ultimately establish limits on the size of the smallest possible magnetic entities useful for technological applications?

The applications focus provided by GMR has helped to stimulate and invigorate the search for new magnetic heterostructures and nanostructures and new magnetoresistive materials. Work on magnetic multilayers is stimulating new thinking about novel devices that can be made by integrating magnetic materials with standard semiconductor technology. Spin-polarized tunneling experiments are helping to elucidate novel magnetic properties as well as demonstrate qualities that have considerable potential for use in devices. The first successes in spin-polarized tunneling between two ferromagnets through an insulating tunnel barrier at room temperature occurred only very recently, with resistance changes of greater than 30 percent now demonstrated.

The continued exponential growth of magnetic storage is by no means assured. As with silicon and optical communications technology, technical and economic limits are looming, and fierce competition is driving the technology toward those limits at a frantic rate. In the case of read-head technology, GMR is now being widely exploited, with CMR and magnetic tunnel junctions as possible follow-ons. In the case of magnetic recording media, the maximum areal density is related to thermal stability, which decreases with bit size. Today's products are believed to be less than a factor of 10 away from this limit. However, the science of magnetism at these scales is still far from complete (see sidebar "Nanocrystals: Building with Artificial Atoms"), and there will undoubtedly be many more innovations in both structure and materials.

NANOCRYSTALS: BUILDING WITH ARTIFICIAL ATOMS

Nanoparticles are nanometer-sized fragments of semiconductors, metals, and dielectrics containing 100 to 1 million atoms. They have captured the attention of researchers and the imagination of the public with their fascinating physical properties. A nanoparticle is often referred to as a nanocrystal or a nanocrystallite when the particle core is a single crystal. The observation of discrete properties resulting from confinement effects in these systems has spawned labels like "quantum dots" and "artificial atoms." Methods of organizing collections of these tailored nanoscale building blocks into new solids that have special optical, electronic, or magnetic properties are being explored. In fact, under the right conditions, these building blocks can assemble themselves (crystallize). When the individual particles are single crystals, the array is often referred to as a nanocrystal superlattice. Their potential value has recently been demonstrated in magnetic recording studies.

The lower left panel shows state-of-the-art CoPtCrB magnetic recording media, while the lower right shows a self-assembled FePt nanocrystal superlattice at the same magnification. Each FePt nanocrystal is 4 nm in diameter. Smaller, more uniform grains in the FePt system will enable more detailed studies of the limits of magnetic recording and the production of ultrahigh-density recording media. The schematic shows the basic combination of organoplatinum, iron carbonyl, and surfactant species employed in the production of FePt nanocrystals, which could provide the means for high-density data storage.

SUMMARY

The ability to compute, communicate, and store information is at the heart of the information revolution. Silicon chip technology, fiber-optic communication technology, and magnetic storage technology are currently the key enabling technologies. All three are experiencing exponential growth in capacity and exponential reductions in cost per operation. Research in condensed-matter and materials physics is the foundation on which these technologies are based. Ultimately these growth rates will flatten as physical and/or economic limits are reached. At that point, either information technology will stop growing or—more likely—wholly new technologies will arise from basic research as the new enablers. In storage, for example, various scanning probe techniques have been proposed as follow-ons to hard disk drives, and computer logic gates made of biological entities or even individual molecules are possible. In optical communications, the use of miniaturized silicon technology or microelectromechanical systems to produce tiny arrays of switches will enable new, low-cost optical networks that do not require translation from optical to electronic and back to optical for regeneration.

Whatever the future may hold, clearly it is the role of basic research to provide the innovations to improve today's technologies and to lay the foundations for new technologies. The critical role of basic research in the technologies of our era was highlighted in a recent address by Alan Greenspan, chairman of the Federal Reserve Board:[1]

> When historians look back at the latter half of the 1990s a decade or two hence, I suspect that they will conclude we are now living through a pivotal period in American economic history. New technologies that evolved from the cumulative innovations of the past half-century have now begun to bring about dramatic changes in the way goods and services are produced and in the way they are distributed to final users. While the process of innovation, of course, is never-ending, the development of the transistor after World War II appears in retrospect to have initiated a special wave of innovative synergies. It brought us the microprocessor, the computer, satellites, and the joining of laser and fiber-optic technologies. By the 1990s, these and a number of lesser but critical innovations had, in turn, fostered an enormous new capacity to capture, analyze, and disseminate information. It is the growing use of information technology throughout the economy that makes the current period unique.

---

[1]Alan Greenspan, "Technology Innovation and Its Economic Impact," address to the National Technology Forum, St. Louis, Mo., April 7, 2000.

Part III

Investing in Our Future:
Priorities and Recommendations

10

A New Era of Discovery

FOUNDATIONS

The Committee's Work

In this volume, *Physics in a New Era: An Overview*, which is the culmination of the National Research Council's six-volume survey *Physics in a New Era*, the committee has considered the science of physics as a whole, reviewing the field broadly and describing its impact on the wider society. This breadth is reflected in the committee's priorities and recommendations, which are meant to sustain and strengthen all of physics in the United States, enabling it to serve important national needs.

- In this chapter, the committee identifies six focused areas of research, which, in its view, are of especially high priority for the discipline as a whole. Some of these important areas coincide with those identified in the earlier volumes of the survey (because they are devoted to various broad areas of physics, those volumes are known as the area volumes[1]). Others are cross-cutting: They reflect new trends in physics, overlap other areas of science, or hold special promise for the development of technology.
- In Chapter 11, the committee recommends improvements in the support for cutting-edge physics research, revisions in physics education, and a strengthening of the role of physics in national security.
- In Chapter 12, the committee makes recommendations on issues of importance to the field as a whole, such as the formation of research partnerships among universities, industry, and national laboratories; the impact on physics of the management policies of the federal science agencies; and the role of information technology in physics.

---

[1]See the preface for a list of the area volumes and the Web site address through which they can be accessed online.

Milestones

The great milestones of 20th-century physics included the discovery of special and general relativity, which revolutionized our view of space and time; the development of quantum mechanics, which provided a roadmap for understanding the subatomic world; the discovery that matter could be transmuted by nuclear fission and fusion, which led to an understanding of the stars; the development of a unified description of the electromagnetic and weak interactions, which was an important step toward a unified theory of the basic forces in nature; the discovery of the Hubble expansion and the cosmic microwave background, which led to an understanding of the birth of the universe; the discovery of new states of matter in superfluids and superconductors, leading to new technologies; the invention of the transistor, the laser, and fiber-optic communication, which gave birth to the information age; and the development of new tools such as x rays, accelerators, and magnetic resonance imaging (MRI), which led to great strides in the biomedical sciences and health care.

These advances and breakthroughs have reshaped all of science and the technology that drives our economy and have opened a new era of discovery. The structure of the physical world may now be probed over distances ranging from 10,000 times smaller than the atomic nucleus to 100,000 times larger than our galaxy. For the first time, phenomena as complex as supernova explosions and weather patterns may be analyzed and understood.

Trends

- New areas in physics are emerging in response to experimental techniques of unprecedented scope and sensitivity as well as the increasing power of computation. We can now control single atoms, observe properties of matter at densities greater than that of the atomic nucleus, design materials with novel properties, study the molecular motors responsible for distributing genetic information during cell division, and probe the earliest moments of the universe.

- At the beginning of the 21st century, physics has become more important for the other sciences, enabling progress in materials science, astronomy, chemistry, geology, and the biomedical sciences. Many of the problems in these areas are increasingly becoming physics-dependent problems; that is, the basic laws of physics play an important role in their understanding.

- Physics is increasingly important for broad technological and eco-nomic development. It has long been a fertile ground for the development of technology—the transistor, the Internet, and MRI are just three examples. The pace of U.S. economic development in the information sciences and other areas will create an increased need for basic physics research in the next decade.
- Physics is now so central to many areas of science and to the solu-tion of problems in health, the environment, and national security that education in physics is more important than ever to advance these areas and to provide a technically trained workforce.
- Physics is increasingly becoming a global enterprise. The rapid exchange of scientific information enabled by the Internet and the excel-lence of physics in Europe and Asia are enough to ensure that. Moreover, the scale of some of the most exciting scientific experiments, on the ground and in space, make international collaboration in science an economic necessity.

Keeping pace with these trends is a challenge for universities, industry, national laboratories, Congress, federal science agencies, and individual scientists. All are key players and all must do their part if physics is to fulfill its promise for the nation in the years ahead.

SCIENTIFIC PRIORITIES AND OPPORTUNITIES

The accomplishments of physics, the increasing power of its instru-ments, and its expanding reach into the other sciences have generated an unprecedented set of scientific opportunities. The committee has described many of them in this volume, and it believes that some are so promising for the decade ahead that their pursuit should be a matter of high national priority. Accordingly, it has identified six "grand challenges," listed below in no particular order. They range across all of physics and overlap other areas of science and engineering. They are selective, some coinciding with the priorities of the area volumes and others cutting more broadly across traditional subfields. Several are of growing importance for technology and economic development. The committee chose them based on their intrinsic scientific importance, their potential for broad impact and application, and their promise for major progress during the decade. In each of the six areas, recent theoretical advances have opened up new questions and set the stage for further synthesis. And in each case, the promise seen for the near future hinges on the emergence of a new generation of instruments providing exquisite precision, great energy and reach, and powerful computational capability. The committee urges that these six high-priority areas be supported strongly by universities, industry, the federal government, and others in the years ahead.

Developing Quantum Technologies

The ability to manipulate individual atoms and molecules will lead to new quantum technologies with applications ranging from the development of new materials to the analysis of the human genome. This ability allows the direct engineering of quantum probabilities, producing novel phenomena such as the presence of many atoms in the same quantum mechanical state with a high probability of spatial overlap and entanglement. Quantum overlap can sometimes extend over distances very large compared to the size of a single atom, as in gaseous Bose-Einstein condensates. A new generation of technology will be developed with construction and operation entirely at the quantum level. Measurement instruments of extraordinary sensitivity, quantum computation, quantum cryptography, and quantum-controlled chemistry are likely possibilities.

Understanding Complex Systems

Theoretical advances and large-scale computer modeling will enable phenomena as complicated as the explosive death of stars and the properties of complex materials to be understood to a degree unimaginable only a few years ago. The rapid advances in massively parallel computing, coupled with equally impressive developments in theoretical analysis, have generated an extraordinary growth in our ability to model and predict complex and nonlinear phenomena and to visualize the results. Problems that may soon be rendered tractable include the strong nuclear force, turbulence and other nonlinear phenomena in fluids and plasmas, the origin of large-scale structure in the universe, and a variety of quantum many-body challenges in condensed-matter, nuclear, atomic, and biological systems. The study of complex systems is inherently of great breadth: Improvements in the understanding of radiation transport, for example, will advance both astrophysics and cancer therapy.

Applying Physics to Biology

Because all essential biological mechanisms ultimately depend on physical interactions between molecules, physics lies at the heart of the most profound insights into biology. Problems central to biology, such as the way molecular chains fold to yield the specific biological properties of proteins, will become accessible to analysis through basic physical laws. Current challenges include the biophysics of cellular electrical activity underlying the functioning of the nervous system, the circulatory system, and the respiratory system; the biomechanics of the motors responsible for all biological movement; and the mechanical and electrical properties of DNA and the enzymes essential for cell division and all cellular processes. Tools developed in physics, particularly for the understanding of highly complex systems, are vital for progress in all these areas. Theoretical approaches developed in physics are being used to understand bioinformatics, biochemical and genetic networks, and computation by the brain.

Creating New Materials

Novel materials will be discovered, understood, and employed widely in science and technology. The discovery of materials such as high-temperature superconductors and new crystalline structures has stimulated new theoretical understanding and led to applications in technology. Several themes and challenges are apparent—the synthesis, processing, and understanding of complex materials composed of more and more elements; the role of molecular geometry and motion in only one or two dimensions; the incorporation of new materials and structures in existing technologies; the development of new techniques for materials synthesis, in which biological processes such as self-assembly can be mimicked; and the control of a variety of poorly understood, nonequilibrium processes (e.g., turbulence, cracks, and adhesion) that affect material properties on scales ranging from the atomic to the macroscopic.

Exploring the Universe

New instruments through which stars, galaxies, dark matter, and the Big Bang can be studied in unprecedented detail will revolutionize our understanding of the universe, its origin, and its destiny. The universe itself is now a laboratory for the exploration of fundamental physics: Recent discoveries have strengthened the connections between the basic forces of nature and the structure and evolution of the universe. New measurements will test the foundations of cosmology and help determine the nature of dark matter and dark energy, which make up 95 percent of the mass-energy of the universe.

Gravitational waves may be directly detected, and the predictions of Einstein's theory for the structure of black holes may be checked against data for the first time. Questions such as the origin of the chemical elements and the nature of extremely energetic cosmic accelerators will be understood more deeply. All of this has given birth to a rich new interplay of physics and astronomy.

Unifying the Forces of Nature

Experiment and theory together will provide a new understanding of the basic constituents of matter. The mystery of the nature of elementary particles deepened in the 1990s with the discovery of the extraordinarily heavy top quark and the observation of oscillations in neutrinos from the Sun and the upper atmosphere, suggesting that neutrinos have extremely tiny masses. During the decade ahead the unknown physics responsible for elementary-particle masses and other properties will begin to reveal itself in experiments at a new generation of high-energy colliders. Possibilities range from the discovery of new and unique elementary particles to more exotic scenarios involving fundamental changes in our description of space and time.

Determining this new physics is an important step toward an historic goal: the discovery of a unified theoretical description of all the fundamental forces of nature—the strong nuclear force, the electroweak forces, and gravity. The most promising and exciting framework for unifying gravity with the other forces is string theory, which proposes that all elementary particles behave like strings at very tiny distances. String theory has also given birth to new and vibrant intersections between physics and pure mathematics. This decade will see much progress toward the goal of discovering a unified theory of the forces of nature.

The scale and complexity of the physics necessary to advance these six priority areas will require increased levels of strategically directed investment and international cooperation. In the United States, the investment must be broad based: from the federal government, from industry, from colleges and universities, and from other supporters of physics research and education. The committee believes that as a result of this focused investment, the decade ahead will see dramatic progress in the above areas and in the many other frontiers described in this overview and in the earlier volumes of the survey. There is little doubt that the new ideas and technologies developed in these quests will enable progress in all the sciences, contribute to the needs of the nation, and benefit the lives of people everywhere.

11

Recommendations I:
Physics and the Wider Society—
Investment, Education, and
National Security

From this survey of physics and its broad impact and the identification of six high-priority scientific opportunities, the committee has formulated a total of nine recommendations. They are designed to strengthen all of physics and to ensure the continued international leadership of the United States. The first five, presented in this chapter, focus on the relationship between physics and the wider society. They address the support of physics by the federal government and the scientific community, physics education, and the role of basic physics research in national security. In each case, the recommendations address problems in need of immediate attention.

INVESTING IN PHYSICS

The character and scope of physics are changing rapidly. There are now extraordinary opportunities for addressing the great questions surrounding the structure of matter, the unification of fundamental forces, and the nature of the universe. New applications to technology and to the life sciences are emerging with increasing frequency. New links are being forged with other key sciences such as chemistry, geology, and astronomy. The increased scope of physics is reflected in the committee's set of scientific priorities and opportunities. Fewer than half concern topics that could be said to belong to the traditional core of physics. The others are new directions branching off from old, with great potential for having a wide impact on science, medicine, national security, and economic growth.

It is widely recognized that the federal government must take primary responsibility for the support of basic research in science, research that is vital for the needs of our nation. Such research is often too broad and distant from commercial development to be a sensible industrial investment. This is particularly true for physics. As a fundamental science, it tends to have a long time lag between discovery in the research lab and impact on the lives of citizens, but by the same token its impact can be all the more profound. Although today's technology and economy are now more closely linked in time with basic physics research than in the past, 10 to 20 years is still the typical interval between a fundamental physics discovery and its impact on society. This can be seen with the laser, magnetic resonance imaging, the optical-fiber transmission line, and many other examples. Much of today's high-tech economy is being driven by the technology that grew out of physics research in the early 1980s.

The unprecedented opportunities facing physics are placing entirely new demands on the field. To study particle interactions at the highest energies, nuclear matter at the highest densities, and the universe at the largest scales and at the earliest moments of its existence requires new instruments of great size and complexity. World-class basic research in quantum phenomena and materials synthesis, with its important economic benefit, can only be carried out with a new generation of sophisticated and precise instrumentation.

Because the federal government plays such a pivotal role in basic physics research, the current level of federal support appears to the committee to be well below the level of support needed to ensure the nation's continuing growth and prosperity. Federal support declined in constant dollars during the 1990s (Figure 11.1). This decline, coming after the modest growth of the 1980s, has meant that growth in the last 20 years averaged only about 2 percent per year. Relative to the size of the economy as measured by the GDP, federal support dropped by more than 20 percent from 1980 to the present (Figure 11.2). This trend, in the view of the committee, has made it more difficult for federal science agencies to support outstanding proposals, making the field of physics less attractive to the excellent people it needs.

Several features in Figure 11.1 are worthy of note. The decrease in the constant-dollar support of basic physics research by the National Science Foundation and the Department of Defense from the early 1980s through 1997 is evident, as is the growth in this support from NASA. Another feature, not always appreciated, is that the federal agency providing the most support for basic research in physics is the Department of Energy. Much of this support is provided through the DOE's national laboratories such as Brookhaven Laboratory, Argonne Laboratory, Fermilab, and the Stanford Linear Accelerator Center. The importance of the DOE to the nation's scientific and technological strength extends far beyond the defense laboratories, discussed in Chapter 8.

Determining the right level for federal support of basic research in any area of science is a very difficult task. However, looking at the economic

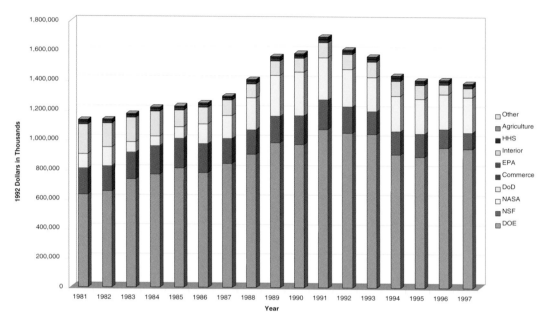

FIGURE 11.1   Federal obligations for basic research in physics by department or agency. Data from the National Science Foundation, *2000 Science and Engineering Indicators*.

benefits now being derived from the physics research of the early 1980s, the committee believes that those years provide an appropriate benchmark for federal investment relative to GDP. It sees no evidence that basic physics research was overfunded then (in fact, federal science agencies were often unable to support high-quality research proposals), and the benefits that have flowed from the research of that era are undeniable.

Along with the high-tech economy, the biological and medical sciences have benefited enormously from basic physics research. It is striking to consider how much of the current biomedical enterprise is driven by the instruments and methods developed in physics through the early 1980s. Such physics-invented technologies as x-ray crystallography, magnetic resonance, fiber optics, electron microscopy, mass spectroscopy, and radioactive tracers are at the heart of the rapid pace of discovery in almost every biomedical laboratory.

The federal government's support of basic research in the life sciences, in contrast with its support of physics research, grew even more rapidly than

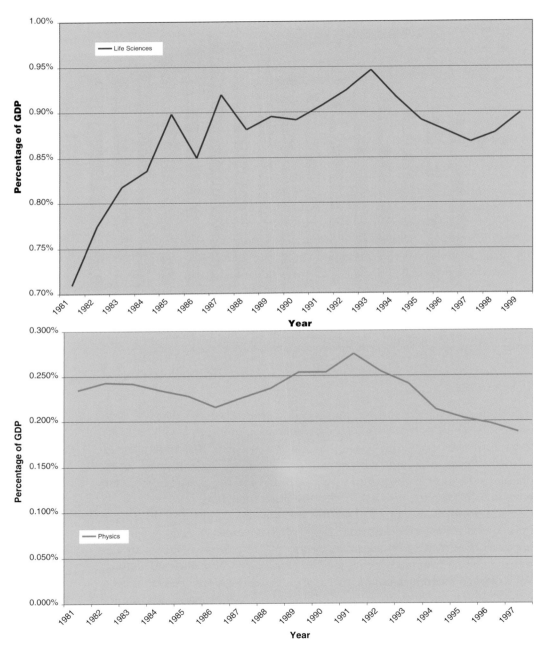

FIGURE 11.2   Federal obligations for basic research in physics and life sciences as percentage of GDP.
SOURCE: Data from the National Science Foundation, *2000 Science and Engineering Indicators.*

the GDP during the past 20 years (Figure 11.2). The committee shares the view of Harold Varmus, past director of the National Institutes of Health, and many other leaders that the discrepancy in funding levels between research in the life sciences and the physical sciences has become so large that future biomedical research will be limited by the lack of new tools and methods that have traditionally been provided to it by chemistry and physics.

> **Recommendation 1. To allow physics to contribute strongly to areas of national need, the federal government and the physics community should develop and implement a strategy for long-term investment in basic physics research. Key considerations in this process should include the overall level of this investment necessary to maintain strong economic growth driven by new physics-based technologies, the needs of other sciences that draw heavily on advances in physics, the expanding scientific opportunities in physics itself, the cost-effectiveness of stable funding for research projects, the characteristic time interval between the investment in basic research and its beneficial impact, and the advantages of diverse funding sources. The Physics Survey Overview Committee believes that to support strong economic growth and provide essential tools and methods for the biomedical sciences in the decade ahead, the federal investment in basic physics research relative to GDP should be restored to the levels of the early 1980s.**

PHYSICS EDUCATION

In both the public and private sectors, it is important that decisions about the development and deployment of a technology be made by people who understand not only its power but also its limitations. Since basic physical principles lie at the heart of this understanding, a first critical goal for our high schools and universities must be scientific literacy—a broad knowledge of these principles on the part of the population at large. A second critical goal is providing the more extensive understanding of physics that is a valuable asset for members of the high-tech workforce. And students must be instilled with an excitement about physics if enough of them are to be drawn into science as a career. Physics education is now failing in each of these critical roles. The present educational system has led to the perception of physics as a difficult subject and not something that well-educated citizens or even members of the technical workforce need to bother with.

Physics departments at colleges and universities need to address these problems on several levels. The curriculum and teaching methods for physics courses must be changed to meet the educational needs of modern society. Current methods and curricula are largely traditional, created to serve a small elite of students planning on careers in physics, narrowly defined. A wider curriculum is needed that embraces broader links with biology, materials science, information technology, and other sciences. The reforms will be difficult, but they are essential if physics education is to foster a widespread understanding of physics and how it applies to the world around us. Numerous alternative and innovative teaching approaches based on research studies have been developed (see Chapter 5, "Physics Education"). They provide successful models for the formulation of more engaging and effective physics teaching methods. Physics departments should also take an active role in the preparation and ongoing training of K-12 teachers of physical science. Although a thorough grounding in the discipline has been widely recognized as playing a key role in excellent K-12 teaching of any physical science, relatively few physical science teachers in the United States have this background.

Recommendation 2. Physics departments should review and revise their curricula to ensure that they are engaging and effective for a wide range of students and that they make connections to other important areas of science and technology. The principal goals of this revision should be (1) to make physics education do a better job of contributing to the scientific literacy of the general public and the training of the technical workforce and (2) to reverse, through a better-conceived, more outward-looking curriculum, the long-term decline in the numbers of U.S. undergraduate and graduate students studying physics. Greater emphasis should also be placed on improving the preparation of K-12 science teachers.

BIG PHYSICS, SMALL PHYSICS

The scale of scientific research is determined by the science itself. Some of the most exciting questions in elementary-particle and nuclear physics, plasma physics, and cosmology, for example, can be answered only with large accelerators and next-generation observatories. The collaboration of many scientists and engineers is essential, the lead time is often long, and the costs can become very high. At the other end of the scale are small table-top experiments or theoretical studies carried out typically by only a few scientists and students. This small scale is appropriate for attacking important problems in condensed-matter and atomic physics, although large facilities for x-ray and neutron scattering are now growing ever more common even here.

The priorities and opportunities the committee describes for physics in the coming decade range over all scales of research. Much of the progress in quantum technologies and the synthesis of new materials will continue to come from single investigators and small groups. To identify the dark matter of the universe and determine the origin of the mass of elementary particles will require large facilities and collaborations of unprecedented size. Unraveling the properties of DNA, RNA, and proteins depends on research ranging from experimentation at large synchrotron light facilities to theoretical modeling and simulation by single investigators. And in many areas, the scale of the research increases as new questions emerge out of old.

Some of the challenges confronting progress in physics are common to research at all scales: for example, the availability of talented people and state-of-the-art instrumentation. But the issues can also be very different depending on the scale of the research. From its consultations and deliberations, the committee identified two areas of concern. One is the very limited availability of adequate funding for single-investigator and small-group research. The other is the critical importance of international planning and priority development when large facilities and large collaborations are involved.

Small Groups and Single Investigators

A large, diverse, and well-supported program of single investigators and small groups is essential for scientifically and technologically important advances and new ideas in physics. Discoveries such as magnetic resonance, the laser, the transistor, and superconductivity have all come out of research carried out by groups of one or two senior scientists often working with a few students. This small-group or single-investigator research is the norm in nearly all of theoretical physics and in those areas that have had a particularly large impact on modern technology, such as biophysics and condensed-matter, atomic, and optical physics. This research environment is particularly well suited to training students because it provides many opportunities for individual creativity and independence. These attractions have led to a large and growing interest in small-group and single-investigator research opportunities among graduate students and to an increasing focus on these areas in the recruiting efforts of many physics departments.

The committee believes that funding for small-group and single-investigator research has become dangerously inadequate and that important opportunities for the nation have been lost as a result. The inadequate funding shows up in numerous ways, including the decreasing success rate in obtaining funding for individual grants and the declining average size, in constant dollars, of individual grants. In many cases the grants have become too small to be viable, and multiple grants are required to sustain a modest research program that formerly required only one. Despite the stiff competition for faculty positions, which has ensured a higher quality of new faculty than ever before, it has become increasingly difficult for young lone investigators working in areas such as condensed-matter physics to obtain federal support.

Because support for small-group and single-investigator research constitutes a small fraction of the total federal investment in physics, this support could be improved substantially with a relatively modest overall increase in funding. The DOE, for example, which provides about two-thirds of the federal support for basic physics research (Figure 11.1), spends more than 80 percent of its budget on physics associated with large facilities and less than 20 percent on small-group and single-investigator physics. Looked at across all federal agencies, support for small groups and single investigators accounts for only about 20 percent of the budget for basic physics research. Increased funding of this research would be a highly cost-effective way to address the technology and technical workforce needs of the future.

Recommendation 3. Federal science agencies should assign a high priority to providing adequate and stable support for small groups and single investigators working at the cutting edge of physics and related disciplines.

Large Facilities and International Collaboration

Many of the most important questions throughout physics can be addressed only with large facilities and the coordinated efforts of many collaborators. The high costs, the long lead times, and the organizational demands lead to many challenges, but the United States must meet them if it is to maintain its leadership role in many of these areas and regain it in others. The necessary steps must begin with the physics communities themselves. As the scale of the research increases, it becomes even more important to assess carefully the scientific opportunities and develop clear priorities nationally and internationally. The divisions of the American Physical Society should strengthen their role in the assessment of opportunities and coordinate the society's efforts with those of corresponding societies in other countries.

The federal funding agencies and their advisory committees responsible for the planning and implementation of large-scale physics research must be connected strongly to the community of physicists in the United States and abroad, with participants serving all of physics rather than representing particular constituencies. Large-scale physics requires extensive R&D, and the federal government must be prepared to support this work well in advance of the start-up of specific facilities. Once initiated, a large-scale project must be managed carefully by the responsible federal agencies and the scientists involved. And with any scientific project, mechanisms must be in place to avoid its continuation beyond its lifetime for forefront research.

When the very largest facilities are involved, such as the next generation of particle colliders necessary to study the high-energy frontier, the planning and the implementation should be international. The federal government should improve its ability to engage in international scientific projects. The Office of Science and Technology Policy can play an important role in this process, working with the agencies supporting large-scale physics research to develop effective protocols for international collaboration, including clear criteria for entrance and exit. A particularly important issue is that of long-term, stable funding commitments. The time scale for large projects can be quite long, and all participating governments must be able to make reliable commitments over the scientific lifetime of the project.

Recommendation 4. While planning and priority setting are important for all of physics, they are especially critical when large facilities and collaborations are necessary. To plan successfully, the community of physicists in the United States and abroad must develop a broadly shared vision and communicate this vision clearly and persuasively. Planning and implementation for the very largest facilities should be international. The federal government should develop effective mechanisms for U.S. participation and leadership in international scientific projects, including clear criteria for entrance and exit.

NATIONAL SECURITY

The agencies of the federal government responsible for the national security of the United States must be able to draw on the highest levels of basic scientific research and expertise.

The Department of Defense supports basic research in physics and other sciences, work that is crucial for the defense interests of the United States. Even with the recent increases, the DOD's support of basic research in physics has declined since the end of the Cold War by approximately 11 percent in constant dollars. In addition, year-to-year fluctuations have made it difficult to maintain important research programs. In the past, DOD laboratories had high-quality programs in basic physics research directly relevant to DOD missions. The people carrying out this research were also able to advise the DOD on physics issues involved in testing, research, and equipment being provided to the DOD by industry. The view of the committee is that over the past decade there has been a substantial decline in the amount and quality of physics research being carried out at DOD laboratories and a corresponding loss of talented people to serve as in-house expert advisors. There is a critical need to ensure that the high-quality physics research and advice required to maintain the technical superiority of our armed forces is being carried out somewhere. The laboratories need to be restored or alternative sources of this expertise must be developed.

The Department of Energy's Office of Defense Programs' national laboratories—Los Alamos, Livermore, and Sandia—have the congressionally mandated duty of verifying the readiness and reliability of the U.S. nuclear arsenal. In the absence of nuclear testing, these laboratories must carry out this duty through a challenging program of component testing and numerical simulation, work that demands the highest quality of scientific personnel, including a vital core of physicists. Many of these researchers were recruited to the laboratories to work on unclassified projects, and they count their ability to participate in basic research as critical to their work.

Security is essential at the laboratories. Lapses can create an adversarial climate between the scientific community and those responsible for security. Low scientist morale and recruitment and retention difficulties now threaten the viability of the laboratories. It is vital to respond to problems in ways that will protect secrecy and yet maintain the creative and scientifically rigorous environment that has been so important to the laboratories throughout their existence.

Recommendation 5. Congress and the Department of Energy should ensure the continued scientific excellence of the Department of Energy's Office of Defense Programs' national laboratories by reestablishing the high priority of long-term basic research in physics and other core competencies important to laboratory missions.

12

Recommendations II: Strengthening Physics Research— Partnerships, Federal Science Agencies, and Physics Information

Any human enterprise must adapt to new opportunities as well as new constraints if it is to stay effective. This is especially true of scientific research, where new opportunities emerge frequently from past accomplishment and technological change and where possibilities depend acutely on available resources. In physics, there was never a time when attention to these changes was more important than it is now. Accordingly, the committee has formulated four additional recommendations focused on the strength of the physics research enterprise. The first deals with the increasingly important role of partnerships among universities, industry, and national laboratories. The next two pertain to the stewardship of federal science agencies. The final recommendation pertains to the rapidly changing role of information technology in physics research and education.

PARTNERSHIPS

Universities are at the core of the U.S. research system, combining the creation of scientific knowledge with its integration and dissemination. Long the envy of the world, they now face serious challenges. The cost of supporting research has grown rapidly in response to expanding scientific opportunities. To provide the faculty and the new infrastructure necessary for research, choices will have to be made and partnerships will have to be strengthened among universities and between universities and national laboratories.

National laboratories are a resource of enormous capability for the United States. They can cross disciplinary boundaries, serve scientists from universities and industry, and address problems of national importance. Realizing the full potential of this resource requires stable funding commitments and effective cooperation with universities and industry. The national laboratories can play an important leadership role in this cooperation and in the formation of research partnerships.

For industry, partnerships with universities and the national laboratories provide important mutual benefits. Industry needs talented graduates to promote growth and innovation, and university faculty and students have much to gain from direct contact with the accelerating pace of technological change. For small companies that may not have R&D capabilities of their own, such partnerships can mean access to unique skills and facilities.

An urgent need is the development of workable intellectual property policies. Agreements must address inventions, patents, copyrights, and the restrictions of proprietary research. Difficulties naturally arise because of the different concerns of the participating institutions. Among large companies, cross-licensing arrangements are common to provide access to important patents. For small companies, exclusivity can make the difference between success and failure. Universities, often the sources of intellectual property but not the commercial users of it, can gain important income from its sale or licensing. National laboratories, which have concerns similar to those of universities, have benefited from legislation authorizing cooperative research and development agreements (CRADAs) for joint research projects with industries. In general, it has been difficult to achieve agreement on terms for research partnerships, particularly since the agreements are often negotiated by people not directly involved in the joint work.

A 1996 report of the Council on Competitiveness, *Endless Frontier, Limited Resources: U.S. R&D Policy for Competitiveness*, noted that "R&D partnerships hold the key to meeting the challenges that our nation now faces." These partnerships include informal collaborations, exchange programs, facility sharing, and various formal relationships such as CRADAs. Technology is creating new modes for partnership, such as long-distance collaboration and information sharing. And partnerships are now more important than ever, since the nation's major industrial laboratories have been forced by divestiture and global economic forces to take a nearer-term, more focused approach to R&D.

Recommendation 6. The federal government, universities and their physics departments, and industry should encourage mutual interactions and partnerships, including industrial liaison programs with universities and national laboratories; visitor programs and adjunct faculty appointments in universities; and university and national laboratory internships and sabbaticals in industry. The

federal government should support these programs by helping to develop protocols for intellectual property issues in cooperative research.

FEDERAL SCIENCE AGENCIES

Program Management

The base programs of the federal science agencies are structured around broad areas of research and are open to a wide range of proposals emerging from the scientific community. These proposals reflect the best ideas and ambitions of scientists throughout the country, and there is no better way for the agencies to ensure the strength and vitality of physics than to respond to the most creative of them. Some of the most exciting proposals are in emerging, interdisciplinary areas. Because the research activities supported by the base programs are diverse and because there is never a guarantee that an individual basic research effort will pay off, it has sometimes been difficult to make the case for the importance of these base programs. By contrast, special funding initiatives providing support for focused areas of research may be more easily understood and have recently been instrumental in securing new support for science agencies. Unless the balance between the base programs and special funding initiatives is monitored carefully, the diversity of high-quality physics research opportunities throughout the country will be jeopardized.

> **Recommendation 7. The federal government should assign a high priority to the broad support of core physics research, providing a healthy balance with special initiatives in focused research directions. Federal science agencies should continue to ensure a foundation that is diverse, evolving, and supportive of promising and creative research.**

Peer Review

The federal government supports scientific research for the broad goal of general societal good and for specific programmatic goals. Predicting which researchers and research projects will best accomplish these goals can never be perfect. The peer review system relies on practicing scientists to advise on the feasibility of proposed research, the competence of the researcher to carry out that work, and the potential impact of the work on science and technology. This system has served our nation well. Its success has convinced funding agencies in many other countries to use it as a model. Many have even gone so far as to make their funding decisions based on reviews of their own country's research programs by U.S. experts familiar with the peer review process in this country.

While the peer review system has worked well, federal science agencies and scientists must always be vigilant for possible abuses of the system. Care should be taken to ensure that reviews are based on scientific merit and are not influenced by other considerations, such as personal connections or prejudices. It is also the case that truly innovative work is often controversial and will naturally generate some mixed reviews. To avoid excessive conservatism, the peer review system must be flexible enough not to demand universally favorable reviews.

Recommendation 8. The peer review advisory process for the allocation of federal government support for scientific research has served our nation well over many decades and is a model worldwide for government investment in research. The peer review process should be maintained as the principal factor in determining how federal research funds are awarded.

PHYSICS INFORMATION

Enabled by greatly increased capabilities in computation and data storage, physics questions are now being addressed by mining data from large databases. For example, in the next few years astrophysical data will be obtained and stored at a rate exceeding several thousand billion bytes per day. Similar data rates are expected from experiments at the new generation of high-energy colliders for nuclear and particle physics. The development of techniques for storing and accessing these enormous repositories of information is becoming essential for basic physics research. As the technical challenges are met, new modalities for data sharing and dissemination of research results become possible.

Concomitant with the rise in importance of large databases containing experimental results has been the increasing reliance on databases for the rapid dissemination of research results. For example, submissions to the archive of physics research based at Los Alamos National Laboratory increased by a factor of nearly 10 over the past decade. The rate of access to this archive is equally impressive, having reached several hundred thousand per day. The archive has become an indispensable tool for conducting research in physics.

At the same time, the American Physical Society has dramatically improved its use of information technology in the publication of research results, establishing important links with the Los Alamos archive. The society is to be commended for reducing the costs of publishing its journals and for supporting the needs of the physics community. It should continue to do so, promoting the use of electronic methods and asserting its preeminent role in the processes of peer review and validation of research results.

Recommendation 9. The federal government, together with the physics community, should develop a coordinated approach for the support of bibliographic and experimental databases and data-mining tools. The use of open standards to foster mutual compatibility of all databases should be stressed. Physicists should be encouraged to make use of these information technology tools for education as well as research. The bibliographic archive based at Los Alamos National Laboratory has played an important role and it should continue to be supported.

Index